Sum of Squares: Theory and Applications

Proceedings of Symposia in
APPLIED MATHEMATICS

Volume 77

Sum of Squares: Theory and Applications

AMS Short Course
Sum of Squares: Theory and Applications
January 14–15, 2019
Baltimore, Maryland

Pablo A. Parrilo
Rekha R. Thomas
Editors

AMERICAN MATHEMATICAL SOCIETY
Providence, Rhode Island

2010 *Mathematics Subject Classification.* Primary 05-XX, 14-XX, 52-XX, 68-XX, 90-XX.

For additional information and updates on this book, visit
www.ams.org/bookpages/psapm-77

Library of Congress Cataloging-in-Publication Data

Library of Congress Cataloging-in-Publication Data

Names: American Mathematical Society. Short Course, Sum of Squares : Theory and Applications (2019 : Baltimore, Maryland), author. — Parrilo, Pablo A., editor. — Thomas, Rekha R., 1967- editor.

Title: Sum of squares : theory and applications : AMS Short Course, Sum of Squares : Theory and Applications, January 14–15, 2019, Baltimore, Maryland / Pablo A. Parrilo, Rekha R. Thomas, editors.

Description: Providence, Rhode Island : American Mathematical Society, [2020] | Series: Proceedings of symposia in applied mathematics, 0160-7634 ; volume 77 | Includes bibliographical references and index.

Identifiers: LCCN 2020012201 | ISBN 9781470450250 (softcover) | ISBN 9781470460402 (ebook)

Subjects: LCSH: Polynomials. | Convex geometry. | Convex sets. | Geometry, Algebraic. | Semidefinite programming. | Mathematical optimization. | AMS: Combinatorics. — Algebraic geometry. | Convex and discrete geometry. | Computer science. | Operations research, mathematical programming.

Classification: LCC QA161.P59 A64 2020 | DDC 511/.6–dc23

LC record available at https://lccn.loc.gov/2020012201

DOI: https://doi.org/10.1090/psapm/077/00677

♾ The paper used in this book is acid-free and falls within the guidelines
established to ensure permanence and durability.
Visit the AMS home page at `https://www.ams.org/`

10 9 8 7 6 5 4 3 2 1 25 24 23 22 21 20

Contents

Preface

This book is a compilation of the lecture notes from the 2019 AMS Short Course on the theory and applications of sum of squares polynomials. Over the past two decades, the theory of nonnegative and sums of squares polynomials has become relevant to several areas of mathematics and related fields. The aim of this book is to showcase some of the recent developments in the field. The introductory chapter gives a brief overview of sums of squares and their connections to real algebraic geometry and polynomial optimization, and also places all chapters in context. The remaining five chapters explore different aspects of sums of squares. The discussion begins with the geometry of spectrahedra, which are the feasible regions of semidefinite programs. This ties in with recent results on expressing convex sets as projections of spectrahedra, a topic further explored in the chapter on lifts of convex sets. The next chapter explains recent generalizations, using classical methods in algebraic geometry, of Hilbert's 1888 theorem cataloging the situations under which all nonnegative polynomials are sums of squares. This is followed by a chapter on recent advances in theoretical computer science that have been obtained by viewing sums of squares as a meta-algorithm for many standard computational tasks. Since sums of squares have found applications in a wide array of practical areas, the book ends with a survey of several examples from engineering, statistics, and operations research.

Pablo A. Parrilo

Rekha R. Thomas

Proceedings of Symposia in Applied Mathematics
Volume **77**, 2020
https://doi.org/10.1090/psapm/077/00680

A brief introduction to sums of squares

Grigoriy Blekherman

Abstract. We give a brief overview and history of nonnegative polynomials and sums of squares. This chapter also provides a guide to how the subsequent chapters connect to this central theory.

This chapter is an informal introduction to the theory of nonnegative polynomials and sums of squares. We assume that the reader is familiar with linear algebra, and especially with positive semidefinite symmetric matrices. For more details on these we recommend [**6**, Appendix A] and [**3**, Section II.12]. The study of nonnegativity and its relation with sums of squares is a classical topic in real algebraic geometry, starting with the work of Hilbert, who classified the cases in terms of degree and number of variables where nonnegative polynomials are always sums of squares of polynomials (Theorem 3.2). Generalizations of this theorem are the subject of Mauricio Velasco's chapter. Hilbert's 17th problem asked whether any globally nonnegative polynomial is a sum of squares of rational functions. Artin's positive solution of Hilbert's 17th problem in the 1920's, using the Artin-Schreier theory of real closed fields left a lasting imprint on the subject.

The study of the subject changed substantially in the early 2000's when it was realized by Lasserre and Parrilo, following earlier work by N.Z. Shor and Choi, Lam and Reznick, that one can use semidefinite programming to efficiently search for sums of squares certificates of nonnegativity in practice [**9**, **14**, **19**, **25**]. The resulting hierarchies (called Lasserre, moment or sums of squares hierarchies) were motivated by theorems of Schmüdgen and Putinar on representations of nonnegative polynomials as sums of squares on compact semialgebraic sets (see Section 5). The sum of squares method has many applications in engineering, robotics and computer science. For more details see Georgina Hall's chapter.

1. Two guiding questions

How can we decide if a function, for instance a polynomial in one variable, takes only nonnegative values? Consider for example, $p(x) = 5x^4 - 4x^3 - x^2 + 2x + 2$. It is not immediately clear, although quite elementary to show, that $p(x)$ is globally nonnegative. However, if we write $p(x)$ as a sum of squares of polynomials:

$$p(x) = (x^2 + 1)^2 + (2x^2 - x - 1)^2,$$

then its nonnegativity is readily apparent.

2010 *Mathematics Subject Classification.* Primary 14P05, 14P10, 90C22.

This simple example immediately poses two important questions:

(1) Are representations of nonnegative polynomials as sums of squares always possible?
(2) If a polynomial is a sum of squares, then how can we find a sum of squares decomposition?

We address the second question first.

2. Finding sum of squares decompositions

Consider a single square $(1-3x+2x^2)^2$. We introduce two vectors, the vector of coefficients $v = (1,-3,2)^T$ and the vector of monomials $\mathbf{x} = (1,x,x^2)^T$. We can write $1-3x+2x^2$ as $v^T\mathbf{x}$ and therefore

$$(1-3x+2x^2)^2 = (\mathbf{x}^Tv)(v^T\mathbf{x}) = \mathbf{x}^T(vv^T)\mathbf{x}.$$

Observe that vv^T is a rank 1 positive semidefinite matrix, and any rank 1 positive semidefinite matrix has the form form vv^T for some real vector v. By applying this procedure to all squares in a sum of squares, we see that any sum of squares $p(x)$ can be written as

$$p(x) = \mathbf{x}^TA\mathbf{x},$$

where A is a positive semidefinite matrix. The converse can be established using the following exercise:

EXERCISE 2.1. Show that any positive semidefinite matrix A can be written as a sum of rank A many positive semidefinite matrices of rank one.

Aside on positive semidefinite matrices: Positive semidefinite matrices form a *convex cone* in the vector space of real symmetric matrices, which we denote by $\mathcal{S}_+^n$. We use notation $A \succeq 0$ to indicate that a symmetric matrix A is positive semidefinite. A good reference for the geometry of $\mathcal{S}_+^n$ is given in [**3**, Section II.12]. A slice of this cone with an affine subspace is called a *spectrahedron*. An interesting spectraherdon, directly related to univariate sums of squares is the *Hankel spectrahedron*, which consists of all positive-semidefinite Hankel matrices (matrices that are constant on anti-diagonals), with $(1,1)$ entry equal to 1. For instance the 3×3 Hankel spectrahedron has the form:

$$H_3 = \left\{ (x,y,z,w)\ :\ \begin{pmatrix} 1 & x & y \\ x & y & z \\ y & z & w \end{pmatrix} \succeq 0 \right\}.$$

We can think of H_3 as a slice of the cone of 3×3 cones of positive semidefinite matrices with a 4-dimensional affine linear subspace given by:

$$\begin{pmatrix} 1 & 0 & 0 \\ 0 & 0 & 0 \\ 0 & 0 & 0 \end{pmatrix} + x\begin{pmatrix} 0 & 1 & 0 \\ 1 & 0 & 0 \\ 0 & 0 & 0 \end{pmatrix} + y\begin{pmatrix} 0 & 0 & 1 \\ 0 & 1 & 0 \\ 1 & 0 & 0 \end{pmatrix} + z\begin{pmatrix} 0 & 0 & 0 \\ 0 & 0 & 1 \\ 0 & 1 & 0 \end{pmatrix} + w\begin{pmatrix} 0 & 0 & 0 \\ 0 & 0 & 0 \\ 0 & 0 & 1 \end{pmatrix}.$$

A matrix is positive semidefinite if and only if all of its principal minors are nonnegative. Therefore the spectrahedron H_3 can be defined by algebraic inequalities, by requiring all principal minors of the above matrix to be nonnegative. For the relation of Hankel spectrahedra to the problem of moments we refer the reader to [**24**]. Cynthia Vinzant's chapter will explain much more about the geometry of $\mathcal{S}_+^n$ and spectrahedra.

We note that Exercise 2.1 says, equivalently, that rank 1 positive semidefinite matrices are precisely *the extreme rays* of $\mathcal{S}_+^n$. Another interesting consequence of our observations is that the minimal number of squares needed to write $p(x)$ as a sum of squares is precisely the minimal rank of any matrix A such that $p(x) = \mathbf{x}^T A\mathbf{x}$. For more information see [**10**]. The logic above can be easily applied to the multivariate setting:

REMARK 2.2 (Multivariate Polynomials). For polynomials in several variables we can a form a vector $\mathbf{x}$ of multivariate monomials up to a certain degree d. For instance, if we want to check that a bivariate polynomial of degree 4 is a sum of squares, the corresponding vector $\mathbf{x}$ is $(1, x, y, x^2, xy, y^2)$. Via the same reasoning, a bivariate polynomial $p(x, y)$ of degree 4 is a sum of squares if and only if there exists a positive semidefinite matrix A such that $p(x, y) = \mathbf{x}^T A\mathbf{x}$. Furthermore, the choice of the monomial basis is not canonical, and a different basis of the vector space of multivariate polynomials of degree at most d may be used to build the vector $\mathbf{x}$.

2.1. Sums of squares and semidefinite programming: Testing whether a given polynomial $p(x)$ is a sum of squares can be done efficiently on a computer using semidefinite programming [**6**, Chapter 3]. We demonstrate this procedure by an example that is small enough to do by hand.

EXAMPLE 2.3. Let $p(x) = 5x^4 - 4x^3 - x^2 + 2x + 2$. By the above, $p(x)$ is a sum of squares if and only if there exists a positive semidefinite matrix A such that

$$p(x) = \mathbf{x}^T A\mathbf{x}, \quad A = \begin{pmatrix} a_{00} & a_{01} & a_{02} \\ a_{01} & a_{11} & a_{12} \\ a_{02} & a_{12} & a_{22} \end{pmatrix}, \quad \mathbf{x} = \begin{pmatrix} 1 \\ x \\ x^2 \end{pmatrix}.$$

A positive semidefinite matrix A such that $p(x) = \mathbf{x}^T A\mathbf{x}$ is called a *Gram matrix* of A. By multiplying $\mathbf{x}^T A\mathbf{x}$ out and comparing coefficients we get equations

$$a_{00} = 2, \ 2a_{01} = 2, \ 2a_{02} + a_{11} = -1, \ 2a_{12} = -4, \ a_{22} = 5.$$

The matrix A therefore has the form:

$$A = \begin{pmatrix} 2 & 1 & a_{02} \\ 1 & -1 - 2a_{02} & -2 \\ a_{02} & -2 & 5 \end{pmatrix}.$$

The choices of a_{02} that make A positive semidefinite describe the *Gram Spectrahedron* of $p(x)$ [**10**]. By considering the determinant of A one can check that A is positive semidefinite for $\frac{1}{4}(1 - \sqrt{185}) \leq a_{02} \leq -1$ and the choices of $a_{02} = -1$ and $a_{02} = \frac{1}{4}(1 - \sqrt{185})$ will lead to rank 2 matrices, i.e. ways of writing $p(x)$ as a sum of two squares. The decomposition from Section 1 has $a_{02} = -1$, and consequently, $a_{11} = 1$.

One can ask whether given a multivariate sum of squares $p(x)$ with rational coefficients there always exists a Gram matrix A filled with rational numbers such that $p(x) = \mathbf{x}^T A\mathbf{x}$? In general, the answer is “no” as shown by Scheiderer in [**22**]. We now discuss semidefinite programming in general.

2.2. A primer on semidefinite programming. Semidefinite programming is a generalization of linear programming where we also allow positive semidefiniteness constraints. More precisely, a general semidefinite program has the form:

$$\begin{aligned} \underset{X}{\text{minimize}} \quad & \langle C, X\rangle \\ \text{subject to} \quad & \langle A_i, X\rangle = b_i \;\; i=1,\dots,m, \\ & X \succeq 0. \end{aligned}$$

The inner product is the trace inner product $\langle A, B\rangle = \operatorname{tr} AB$. We note that if X is restricted to being a diagonal matrix then the above semidefinite program becomes a linear program, since positive-semidefiniteness constraint only ensures that the diagonal entries are nonnegative.

Semidefinite programs can be efficiently solved using interior point methods [**4**]. However, these methods will struggle when the matrix size grows large, and this becomes quite relevant for sums of squares methods, as seen in the remark below.

REMARK 2.4. Observe that the size of the Gram matrix A is given by the number of relevant monomials (or more precisely the dimension of the vector space from which squares come). If we want to check whether an n-variate polynomial of degree $2d$ is a sum of squares, then $\mathbf{x}$ should be the vector of all monomials in n variables of degree at most d, and therefore the size of A is $\binom{n+d}{d} \times \binom{n+d}{d}$. As the number of variables and the degree grows the size of A increases rapidly, and this is perhaps the greatest practical limitation of the sum of squares method, as matrices may soon become too large to be effectively handled by SDP solvers.

2.3. Lifted representations. We can define a map Φ which sends a symmetric matrix A to the polynomial $\mathbf{x}^T A\mathbf{x}$. Another way of interpreting and generalizing Example 2.3 is that the map Φ projects the cone $\mathcal{S}^n_+$ to sums of squares, or equivalently, it provides a *lifted SDP representation* of the cone of sums of squares. For instance, Example 2.3 shows that a univariate polynomial $a_0+a_1x+a_2x^2+a_3x^3+a_4x^4$ of degree 4 is a sum of squares if and only if there exists a positive semidefinite matrix

$$\begin{pmatrix} a_0 & \frac{1}{2}a_1 & b \\ \frac{1}{2}a_1 & c & \frac{1}{2}a_3 \\ b & \frac{1}{2}a_3 & a_4 \end{pmatrix} \succeq 0,$$

where $2b + c = a_2$. Therefore the cone of univariate sums of squares of degree 4 is a projection of the cone S^3_+ of 3×3 positive semidefinite matrices.

One can study, in general, which sets can be written as projections of "simple sets", for instance spectrahedra. Such lifted representations allow to us to lift an optimization problem from a complicated set to the simple set upstairs, where optimization can be done efficiently. See Hamza Fawzi's chapter for more on lifted representations.

3. Nonnegative polynomials and sums of squares

We now discuss in detail the comparison between nonnegative polynomials and sums of squares. The story begins in 1885 at the Ph.D. defense of Hermann Minkowski, where David Hilbert was one of the examiners. During his defense Minkowski claimed that there exist nonnegative polynomials that are not sums of squares, although he did not provide an example or a proof. Three years later

Hilbert published a seminal paper [**12**] classifying all cases in terms of number of variables n and degree $2d$ for which all n-variate nonnegative polynomials of degree at most $2d$ are sums of squares.

We use $P_{n,\leq 2d}$ and $\Sigma_{n,\leq 2d}$ to respectively denote the sets of n-variate nonnegative polynomials and sum of squares of degree at most $2d$. We note that both form convex cones in the vector space $\mathbb{R}[x]_{n,\leq 2d}$ of n-variate polynomials of degree at most $2d$, and $\dim \mathbb{R}[x]_{n,\leq 2d} = \binom{n+2d}{2d}$.

EXERCISE 3.1. Show that $P_{n,\leq 2d}$ and $\Sigma_{n,\leq 2d}$ are closed full-dimensional convex cones with no lines in $\mathbb{R}[x]_{n,\leq 2d}$. For a hint see [**6**, Exercise 4.17].

We now state Hilbert's 1888 theorem:

THEOREM 3.2. $P_{n,\leq 2d} = \Sigma_{n,\leq 2d}$ *only in the following three cases:*

(1) $n = 1$, *the case of univariate polynomials*
(2) $2d = 2$, *the case of quadratic polynomials*
(3) $n = 2$, $2d = 4$, *the exceptional case of ternary quartics.*

In all other cases there exist nonnegative polynomials that are not sums of squares.

It should be noted that Hilbert did not provide an explicit nonnegative polynomial that is not a sum of squares of polynomials. The first explicit example appeared only much later and is due to Motzkin [**16**] (See Exercise 3.5). The main difficulty of constructing an example is certifying that a given polynomial is globally nonnegative without relying on a sums of squares decomposition. Motzkin's idea was to guarantee nonnegativity via classical inequalities! The Motzkin polynomial

$$M(x, y) = x^2y^4 + x^4y^2 + 1 - 3x^2y^2$$

is nonnegative by applying the Arithmetic Mean/Geometric Mean inequality, and an argument detailed in the exercises below shows that it is not a sum of squares.

EXERCISE 3.3. The *Newton Polytope* N_p of a polynomial p is the convex hull of the vectors of monomial exponents that occur in p. For example, the Newton Polytope of x^2+y+1 is the convex hull of vectors $(2, 0), (0, 1)$ and $(0, 0)$. Show that if a polynomial $p = \sum_i q_i^2$ is a sum of squares, then the Newton Polytope of each q_i is contained in $\frac{1}{2}N_p$. Hint: It helps to consider the convex hull of the Newton polytopes of q_i's.

EXERCISE 3.4. Define homogenization $\tilde{p}$ of a polynomial $p(x_1, \ldots, x_n)$ of degree d by introducing a new variable x_0 and multiplying all monomials in p by a power of x_0, so that all monomials have degree d. More formally:

$$\tilde{p} = x_0^d \cdot p\left(\frac{x_1}{x_0}, \ldots, \frac{x_n}{x_0}\right).$$

Show that p is nonnegative if and only if $\tilde{p}$ is, and the same is also true for sums of squares.

In view of the above exercise, questions about global nonnegativity and sums of squares are often studied for homogeneous polynomials, which are also called forms.

EXERCISE 3.5. Use Exercise 3.3 to show that the Motzkin polynomial is not a sum of squares. Also show that the homogeneous polynomial $S(x, y, z) = x^4y^2 + y^4z^2 + z^4x^2 - 3x^2y^2z^2$, constructed by Choi and Lam, is nonnegative but not a sum of squares.

Since then many other examples of nonnegative polynomials that are not sums of squares have appeared, particularly in the work of Choi, Lam and Reznick (see [**21**] for a nice overview).

After proving his 1888 Theorem, Hilbert showed in 1893 that any nonnegative bivariate polynomial is a sum of squares of rational functions. At first, writing

$$f = \sum \left(\frac{p_i}{q_i}\right)^2$$

may seem quite limiting, however observe that if we bring all of the rational functions to a common denominator r then we are writing

$$f \cdot r^2 = \sum s_i^2,$$

i.e. we hope to find a square multiplier r^2, such that $f \cdot r^2$ is a sum of squares of polynomials. This is clearly a generalization of writing f as a sum of squares of polynomials, where we take r to be the constant polynomial 1. In fact we can even allow sum of squares multipliers, instead of just a single square:

EXERCISE 3.6. Show that f is a sum of squares of rational functions if and only if there exists a sum of squares polynomial h such that $f \cdot h$ is a sum of squares of polynomials.

Hilbert's 17th problem asked to prove that a nonnegative polynomial in any number of variables is a sum of squares of rational functions. Artin's original proof, which relied on the newly established Artin-Schreier theory of real closed fields, was purely existential. In particular, it did not provide any bounds on the degree of the multiplier h [**8**]. Searching for multiplier h of fixed degree such that $f \cdot h$ is a sum of squares can also be formulated as a semidefinite program, so it is important for computations to understand in what degree rational certificates exist.

To this day, our understanding of both upper and lower degree bounds is quite poor. For the best current upper bound see [**15**] and for the lower bounds see [**5**]. For the special case of nonnegative polynomials on curves see [**7**]. While philosophically, Artin's solution to Hilbert's 17th problem provides a quite satisfactory answer on the power of sums of squares to explain nonnegativity, computationally much remains unknown. As we are about to see, in applications we usually just use sums of squares of polynomials, as rational sums of squares often lead to numerical problems in solving semidefinite programs.

4. Sums of squares and optimization

We briefly discuss the relevance of sums of squares in optimization. We begin with a crucial observation, that being able to optimize, i.e. minimize or maximize, polynomials efficiently is equivalent to algorithmically understanding nonnegative polynomials. Consider the problem of finding the global infimum of a polynomial $p(x)$:

$$\lambda = \inf_{x \in \mathbb{R}^n} p(x).$$

Then λ is the best lower bound for p, or equivalently, λ is the largest value that we can subtract so that $p(x) - \lambda$ is nonnegative. Therefore, if we can efficiently test whether a polynomial is nonnegative, then we can do optimization.

Unfortunately, the cone of nonnegative polynomials is quite complicated, and testing nonnegativity of a polynomial is already an NP-complete problem when the degree is 4. However, we can replace nonnegative polynomials, with "obviously nonnegative" polynomials, i.e. sums of squares! What we can compute, in practice, is instead λ^*, which is the largest value, such that $p(x) - \lambda^*$ is a sum of squares. It is clear that λ^* is a lower bound for the true infimum λ, and furthermore, very importantly, the fact that $p(x) - \lambda^*$ is nonnegative comes with a *sum of squares certificate.*

In applications, we often have problems with additional constraints, but the idea behind sum of squares approach is the same: we "relax" the intractable set of nonnegative polynomials with the tractable set of "obviously nonnegative" polynomials, which we construct using sums of squares. More details are given in the next section. This simple idea has had a profound impact in engineering and theoretical computer science, see [**1**, **2**, **13**] for some examples. For applications of sums of squares method in optimization see Georgina Hall's chapter and for applications in computer science see Ankur Moitra's chapter.

5. Adding constraints

Suppose that instead of global nonnegativity we want to understand polynomials nonnegative on an algebraic set X in $\mathbb{R}^n$ defined by equations $g_1(x) = \cdots = g_k(x) = 0$. Such an algebraic set is also called *a variety.* Since all polynomials g_i are identically zero on X, we have a larger class of "obviously nonnegative" polynomials on X, namely sums of squares and also multiples of g_i, which we can multiply and add together. This leads to an interesting phenomenon of degree cancellation, which we illustrate by an example:

EXAMPLE 5.1. Let $X = \{0, 1\} \subset \mathbb{R}$ be given by $g(x) = x(1-x) = 0$. We want to certify that the function x is nonnegative on X. This is not possible using sums of squares only, since the degree of x is one. However we can write:

$$x = x^2 + x(1-x),$$

so x is "obviously nonnegative" on X.

By considering sums of squares on varieties it is possible to generalize Hilbert's 1888 theorem, and find more cases of equality between nonnegative polynomials and sums of squares, even without degree cancellation. See Mauricio Velasco's chapter for more details.

This brings us to a very important theorem, which has been crucial for the development of applications of sums of squares in optimization. If we are allowed to use degree cancellation, then is it true that any polynomial p nonnegative on *a compact* variety X is "obviously nonnegative"? This is false in general (see Exercise 5.5 below), but it is true for strictly positive p.

THEOREM 5.2 (Schmüdgen, [**23**]). *Let $X \subset \mathbb{R}^n$ be a compact real variety defined by equations $g_1(x) = \cdots = g_k(x) = 0$ and let f be a polynomial strictly positive on X. Then f is "obviously nonnegative" on X, i.e. there exist polynomials $h_1, \ldots, h_k$ and a sum of squares σ such that*

$$f = \sigma + g_1 h_1 + \cdots + g_k h_k.$$

Due to the central role that Schmüdgen's theorem plays in applications in optimization, we will discuss its statement and applications in detail.

Degree Truncation: To make Schmüdgen's Theorem useful for computations we *search for certificates of bounded degree*. This means given a degree bound d we search for $h_1, \dots, h_k$ and a sum of squares σ such that

$$\deg g_i h_i \le d \quad \text{for} \quad 1 \le i \le k \quad \text{and} \quad \deg \sigma \le d.$$

This can still be formulated as a semidefinite programming problem, and is referred to in the literature as *Lasserre* or *Lasserre-Parrilo* or *sums of squares* or *moment relaxation* (technically moment relaxation refers to the dual semidefinite problem).

EXAMPLE 5.3. Let X be the set $\{-1, 0, 1\} \subset \mathbb{R}$ given by the equation $g(x) = x(x^2-1)$, and let $p(x) = 2 + x$. It is clear that p is nonnegative on X. If we choose two as a bound for the degree of an "obviously nonnegative" representation, then p is cannot have a certificate, since we cannot use any multiples of g, and p is not a sum of squares, since it is not globally nonnegative. We leave it as an exercise to show that we can show that p is obviously nonnegative with degree bound 4.

Low Degree Certificates: The size of the semidefinite program that needs to be solved depends on the degree truncation d. Usually, the size of underlying matrices grows quickly with d (cf. Remark 2.4). Therefore, for practical consideration and from point of view of computational complexity, understanding of *low-degree certificates* is paramount. See Ankur Moitra's chapter for the theoretical computer science perspective.

Semialgebraic sets: We stated Schmüdgen's Positivestellensatz only for varieties, but it also holds for semialgebraic sets given by inequalities $g_i \ge 0$ (such sets are called basic closed semialgebraic sets). In this case products of g_i are obviously nonnegative, as are sums of squares. An "obviously nonnegative" polynomial on X has the form

$$\sum_j \sigma_j \Pi_{a_i \in \{0,1\}} g_i^{a_i},$$

where σ_j are sums of squares. By Schmüdgen's theorem any strictly positive polynomial on X is "obviously nonnegative". Under a mild technical assumption Putinar's Positivstellensatz [**20**] guarantees that we do not need to consider products with more than one g_i in them, and we can take "obviously nonnegative" polynomials of the form:

$$\sigma_0 + \sum_j \sigma_j g_j.$$

Compactness is necessary: Schmüdgen's theorem may fail if the variety X is not compact. For instance consider the case of global nonnegativity, when we have no constraints and the variety X is $\mathbb{R}^n$. The following Exercise is useful:

EXERCISE 5.4. Show that if a polynomial p of degree $2d$ is a sum of squares, then the degree $2d$ part of p, i.e. the sum of all monomials of degree exactly $2d$ with the coefficients from p, is also a sum of squares. Hint: Use Exercise 3.3.

Consider again the Motzkin form $M(x, y, z) = x^2y^4 + x^4y^2 + z^6 - 3x^2y^2z^2$ (or any homogeneous polynomial that is not a sum of squares). It is now easy to show that $M(x, y, z) + 1$ is strictly positive on $\mathbb{R}^n$ but is not "obviously nonnegative", since $M(x, y, z)$ is not a sum of squares.

Strict positivity is necessary: If f is only nonnegative on X and is not strictly positive, then it may fail to be "obviously nonnegative". The following example is taken (with minimal modification) from [**11**]. Let $X \subset \mathbb{R}^2$ be the curve defined by the equation $g(x) = x^4 - x^3 + y^2 = 0$. Then X is a compact subset of $\mathbb{R}^2$ with a singularity at the origin.

EXERCISE 5.5. Show that the function $f(x) = x$ is nonnegative on X, but f is not "obviously nonnegative".

There has been extensive work on additional conditions under which nonnegative polynomials are guaranteed to be "obviously nonnegative" [**17, 18**]. This is sometimes referred to as "finite convergence" of the sum of squares hierarchy.

References

[1] Boaz Barak, Fernando G. S. L. Brandão, Aram W. Harrow, Jonathan Kelner, David Steurer, and Yuan Zhou, *Hypercontractivity, sum-of-squares proofs, and their applications [extended abstract]*, STOC'12—Proceedings of the 2012 ACM Symposium on Theory of Computing, ACM, New York, 2012, pp. 307–326, DOI 10.1145/2213977.2214006. MR2961513

[2] Boaz Barak, Jonathan A. Kelner, and David Steurer, *Dictionary learning and tensor decomposition via the sum-of-squares method [extended abstract]*, STOC'15—Proceedings of the 2015 ACM Symposium on Theory of Computing, ACM, New York, 2015, pp. 143–151, DOI 10.1145/2746539.2746605. MR3388192

[3] Alexander Barvinok, *A course in convexity*, Graduate Studies in Mathematics, vol. 54, American Mathematical Society, Providence, RI, 2002. MR1940576

[4] Aharon Ben-Tal and Arkadi Nemirovski, *Lectures on modern convex optimization: Analysis, algorithms, and engineering applications*, MPS/SIAM Series on Optimization, Society for Industrial and Applied Mathematics (SIAM), Philadelphia, PA; Mathematical Programming Society (MPS), Philadelphia, PA, 2001. MR1857264

[5] Grigoriy Blekherman, João Gouveia, and James Pfeiffer, *Sums of squares on the hypercube*, Math. Z. **284** (2016), no. 1-2, 41–54, DOI 10.1007/s00209-016-1644-7. MR3545483

[6] G. Blekherman, P. A. Parrilo, and R. R. Thomas, *Semidefinite optimization and convex algebraic geometry*, SIAM, 2012.

[7] Grigoriy Blekherman, Gregory G. Smith, and Mauricio Velasco, *Sharp degree bounds for sum-of-squares certificates on projective curves* (English, with English and French summaries), J. Math. Pures Appl. (9) **129** (2019), 61–86, DOI 10.1016/j.matpur.2018.12.010. MR3998790

[8] Jacek Bochnak, Michel Coste, and Marie-Françoise Roy, *Real algebraic geometry*, Ergebnisse der Mathematik und ihrer Grenzgebiete (3) [Results in Mathematics and Related Areas (3)], vol. 36, Springer-Verlag, Berlin, 1998. Translated from the 1987 French original; Revised by the authors. MR1659509

[9] M. D. Choi, T. Y. Lam, and B. Reznick, *Sums of squares of real polynomials*, K-theory and algebraic geometry: connections with quadratic forms and division algebras (Santa Barbara, CA, 1992), Proc. Sympos. Pure Math., vol. 58, Amer. Math. Soc., Providence, RI, 1995, pp. 103–126. MR1327293

[10] Lynn Chua, Daniel Plaumann, Rainer Sinn, and Cynthia Vinzant, *Gram spectrahedra*, Ordered algebraic structures and related topics, Contemp. Math., vol. 697, Amer. Math. Soc., Providence, RI, 2017, pp. 81–105, DOI 10.1090/conm/697/14047. MR3716067

[11] João Gouveia and Tim Netzer, *Positive polynomials and projections of spectrahedra*, SIAM J. Optim. **21** (2011), no. 3, 960–976, DOI 10.1137/100801913. MR2837559

[12] David Hilbert, *Ueber die Darstellung definiter Formen als Summe von Formenquadraten* (German), Math. Ann. **32** (1888), no. 3, 342–350, DOI 10.1007/BF01443605. MR1510517

[13] Samuel B. Hopkins, Jonathan Shi, Tselil Schramm, and David Steurer, *Fast spectral algorithms from sum-of-squares proofs: tensor decomposition and planted sparse vectors*, STOC'16—Proceedings of the 48th Annual ACM SIGACT Symposium on Theory of Computing, ACM, New York, 2016, pp. 178–191, DOI 10.1145/2897518.2897529. MR3536564

[14] Jean B. Lasserre, *Global optimization with polynomials and the problem of moments*, SIAM J. Optim. **11** (2000/01), no. 3, 796–817, DOI 10.1137/S1052623400366802. MR1814045

[15] H. Lombardi, D. Perrucci, and M.-F. Roy, *An elementary recursive bound for effective positivstellensatz and hilbert 17-th problem*, arXiv preprint `arXiv:1404.2338`, 2014.

[16] T. S. Motzkin, *The arithmetic-geometric inequality*, Inequalities (Proc. Sympos. Wright-Patterson Air Force Base, Ohio, 1965), Academic Press, New York, 1967, pp. 205–224. MR0223521

[17] Jiawang Nie, *Certifying convergence of Lasserre's hierarchy via flat truncation*, Math. Program. **142** (2013), no. 1-2, Ser. A, 485–510, DOI 10.1007/s10107-012-0589-9. MR3127083

[18] Jiawang Nie, *Optimality conditions and finite convergence of Lasserre's hierarchy*, Math. Program. **146** (2014), no. 1-2, Ser. A, 97–121, DOI 10.1007/s10107-013-0680-x. MR3232610

[19] P. A. Parrilo. *Structured semidefinite programs and semialgebraic geometry methods in robustness and optimization.* PhD thesis, California Institute of Technology, 2000.

[20] Mihai Putinar, *Positive polynomials on compact semi-algebraic sets*, Indiana Univ. Math. J. **42** (1993), no. 3, 969–984, DOI 10.1512/iumj.1993.42.42045. MR1254128

[21] Bruce Reznick, *Some concrete aspects of Hilbert's 17th Problem*, Real algebraic geometry and ordered structures (Baton Rouge, LA, 1996), Contemp. Math., vol. 253, Amer. Math. Soc., Providence, RI, 2000, pp. 251–272, DOI 10.1090/conm/253/03936. MR1747589

[22] Claus Scheiderer, *Sums of squares of polynomials with rational coefficients*, J. Eur. Math. Soc. (JEMS) **18** (2016), no. 7, 1495–1513, DOI 10.4171/JEMS/620. MR3506605

[23] Konrad Schmüdgen, *The K-moment problem for compact semi-algebraic sets*, Math. Ann. **289** (1991), no. 2, 203–206, DOI 10.1007/BF01446568. MR1092173

[24] Konrad Schmüdgen, *The moment problem*, Graduate Texts in Mathematics, vol. 277, Springer, Cham, 2017. MR3729411

[25] N. Z. Shor, *Class of global minimum bounds of polynomial functions*, Cybernetics and Systems Analysis, **23** (6):731–734, 1987.

SCHOOL OF MATHEMATICS, GEORGIA INSTITUTE OF TECHNOLOGY, 686 CHERRY STREET, ATLANTA, GA 30332-0160 USA

Email address: `greg@math.gatech.edu`

Proceedings of Symposia in Applied Mathematics
Volume **77**, 2020
https://doi.org/10.1090/psapm/077/00681

The geometry of spectrahedra

Cynthia Vinzant

ABSTRACT. Spectrahedra are convex, semialgebraic sets that are fundamental objects in the theory of semidefinite programming, and more generally matrix theory, convex optimization, and real algebraic geometry. Here we develop some of the relevant background from real algebraic and convex geometry, and explore its impacts on the geometry of spectrahedra and applications to cones of sums of squares and moments.

1. Introduction

A spectrahedron is the intersection of the cone of positive semidefinite matrices with an affine linear space. These appear in convex optimization as the feasible sets of semidefinite programs. In what follows, we investigate the geometry of spectrahedra using tools from real algebraic geometry and convexity and discuss implications for spectrahedra related to sums of squares. The geometry of spectrahedra depends heavily on the geometry of the cone of positive semidefinite matrices.

DEFINITION 1.1. A real symmetric matrix A is *positive definite*, denoted $A \succ 0$, if all of its eigenvalues are positive. The matrix A is *positive semidefinite*, denoted $A \succeq 0$, if all of its eigenvalues are nonnegative.

There are several equivalent characterizations of positive semidefinite-ness. For an $N \times N$ real symmetric matrix A, the following conditions are equivalent:

(i) all eigenvalues of A are nonnegative,
(ii) all principal minors of A, $\det(A_{I,I})$ for $I \subseteq \{1, \ldots, N\}$, are nonnegative,
(iii) $\mathbf{v}^T A \mathbf{v} \geq 0$ for all $v \in \mathbb{R}^N$, and
(iv) there exists a matrix $B \in \mathbb{R}^{N \times k}$ where $k = \text{rank}(A)$ and

$$A \;=\; BB^T \;=\; (\langle \mathbf{r}_i, \mathbf{r}_j \rangle)_{1 \leq i,j \leq N} \;=\; \sum_{i=1}^{k} \mathbf{c}_i \mathbf{c}_i^T$$

where $\mathbf{r}_1, \ldots, \mathbf{r}_N$, $\mathbf{c}_1, \ldots, \mathbf{c}_k$ are the rows and columns of B.

Let $\mathcal{S}^N$ denote the vector space of real symmetric $N \times N$ matrices and $\mathcal{S}_+^N$ denote the set of $N \times N$ positive semidefinite matrices. The set $\mathcal{S}_+^N$ is a full-dimensional *convex semialgebraic cone* in the real vector space $\mathcal{S}^N \cong \mathbb{R}^{\binom{N+1}{2}}$. General convex

---

2010 *Mathematics Subject Classification.* Primary 90C22, 15B48, 14P10, 52A41.
Key words and phrases. Spectrahedra, positive semidefinite matrix, semidefinite program, sums of squares, moment problems.

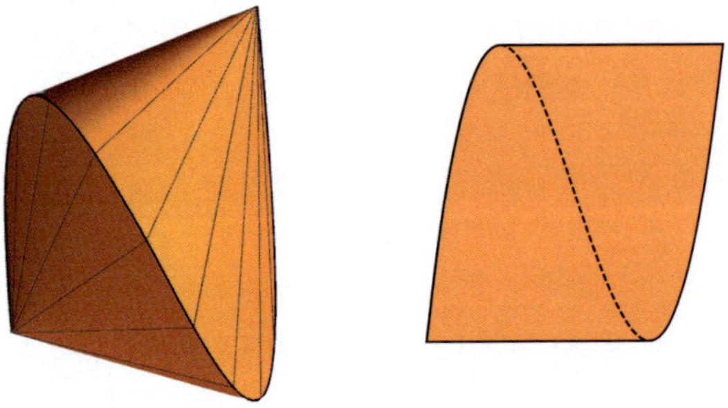

FIGURE 1. The spectrahedron in $\mathcal{S}^4$ defined by (1.1) and its image under the coordinate projection $(x, y, z) \mapsto (x, z)$.

and semialgebraic sets will be defined and discussed in Section 2. One example is the intersection of $\mathcal{S}_+^N$ with an affine linear subspace of $\mathcal{S}^N$.

DEFINITION 1.2. A *spectrahedron* is the intersection of the cone of positive semidefinite matrices with an affine linear space. The affine linear space can be parametrized as $\mathcal{L} = \{A(\mathbf{x}) : \mathbf{x} \in \mathbb{R}^m\}$ where $A(\mathbf{x}) = A_0 + \sum_{i=1}^m x_i A_i$ is a linear matrix pencil given by $A_0, \dots, A_m \in \mathcal{S}^N$. This linearly identifies the spectrahedron with

$$\mathcal{L} \cap \mathcal{S}_+^N \;\cong\; \left\{ \mathbf{x} \in \mathbb{R}^m \;:\; A(\mathbf{x}) = A_0 + \sum_{i=1}^m x_i A_i \succeq 0 \right\}.$$

We will refer to both sets (and any set linearly isomorphic to them) as spectrahedra.

EXAMPLE 1.3. Consider the three-dimensional affine linear space parametrized

$$A(x,y,z) \;=\; A_0 + xA_1 + yA_2 + zA_3 \;=\; \begin{pmatrix} 1 & x & y & z \\ x & 1 & x & y \\ y & x & 1 & x \\ z & y & x & 1 \end{pmatrix} \tag{1.1}$$

where (x, y, z) runs over $\mathbb{R}^3$ and the matrices $A_0, A_1, A_2, A_3 \in \mathcal{S}^4$ have the form $(A_k)_{ij} = 1$ if $|i - j| = k$ and 0 otherwise. The intersection of this affine space with the cone of 4×4 positive semidefinite matrices $\mathcal{S}_+^4$ gives the spectrahedron $S = \{(x, y, z) \in \mathbb{R}^3 : A(x, y, z) \succeq 0\}$ shown on the left in Figure 1. ◇

EXAMPLE 1.4. Another important example of a spectrahedron is any polyhedron. A *polyhedron* is the set of points satisfying finitely many affine linear inequalities. That is, it is a set that can be written as

$$P = \left\{ \mathbf{x} \in \mathbb{R}^m : a_{0j} + \sum_{i=1}^m a_{ij} x_i \geq 0, \; j = 1, \dots, N \right\},$$

where $a_{ij} \in \mathbb{R}$ for all $i = 0, \dots, m$ and $j = 1, \dots, N$. We can represent P as a spectrahedron using diagonal matrices. Indeed, for each $i = 0, \dots, m$, let A_i be the $N \times N$ diagonal matrix with diagonal entries $(a_{i1}, \dots, a_{iN})$. For any $\mathbf{x} \in \mathbb{R}^m$, the linear combination $A_0 + \sum_{i=1}^m A_i x_i$ is again a diagonal matrix, which is positive semidefinite if and only if its diagonal entries are nonnegative. This writes P as the spectrahedron,

$$P = \left\{ \mathbf{x} \in \mathbb{R}^m : A_0 + \sum_{i=1}^{m} A_i x_i \succeq 0 \right\}.$$

◇

Another important class of convex semialgebraic sets are *spectrahedral shadows*, which are the image of spectrahedra under linear projections. One can define these as sets that can be written in the form

$$\left\{ (x_1, \dots, x_m) \in \mathbb{R}^m : \exists (y_1, \dots, y_n) \in \mathbb{R}^n \text{ with } A_0 + \sum_{i=1}^{m} x_i A_i + \sum_{j=1}^{n} y_j B_j \succeq 0 \right\}$$

for some matrices $A_i, B_j \in \mathcal{S}^N$. For example, the projection of the spectrahedron S in Example 1.3 onto the coordinates (x, z) is shown on the right in Figure 1. As we will see, spectrahedral shadows are not always spectrahedra. This is a major distinction with polyhedra – the image of a polyhedron under linear projection is always a polyhedron.

Like the cone of positive semidefinite matrices, spectrahedra and spectrahedral shadows are convex, semialgebraic sets, and we can analyze both their algebraic and convex structure.

In what follows we build up some of the necessary background from real algebraic geometry and convexity (Section 2), introduce convex duality with a focus on spectrahedral cones (Section 3), examine the facial structure of the cone of positive semidefinite matrices and spectrahedra (Section 4), and end by discussing applications of this theory to sums of squares and nonnegative polynomials (Section 5).

2. Background

Throughout, we use the standard Euclidean inner product on $\mathbb{R}^n$, where for two vectors $\mathbf{x} = (x_1, \dots, x_n)$ and $\mathbf{y} = (y_1, \dots, y_n)$, $\langle \mathbf{x}, \mathbf{y} \rangle = \sum_{i=1}^n x_i y_i$. On $\mathcal{S}^N$, we use the trace inner product $\langle A, B \rangle = \text{trace}(AB)$.

2.1. Real algebraic geometry and semialgebraic sets. The fundamental building blocks of real algebraic geometry are *semialgebraic sets*, which are sets defined by polynomial equations and inequalities. For more details on the subject, see [**3, 10**]. Formally, we have the following:

DEFINITION 2.1. A *basic closed semialgebraic set* in $\mathbb{R}^n$ is a set of the form

$$\{\mathbf{p} \in \mathbb{R}^n : g_1(\mathbf{p}) \geq 0, \dots, g_s(\mathbf{p}) \geq 0\}$$

where $g_1, \dots, g_s \in \mathbb{R}[x_1, \dots, x_n]$ are polynomials. A *semialgebraic set* is one that can be written as a finite boolean combination of basic closed semialgebraic sets, that is, one built from a finite collection of basic closed semialgebraic sets using finitely many complements, intersections, and unions.

EXAMPLE 2.2. A disk and a square are both basic closed semialgebraic subsets of $\mathbb{R}^2$. Their union will be a closed semialgebraic set, but might not be a basic one. If D is the unit disk defined by $x^2 + y^2 \leq 1$ and S is the square defined by $0 \leq x \leq 2$ and $-1 \leq y \leq 1$, then the union $D \cup S$ is semialgebraic, but not basic. These are shown is Figure 2.

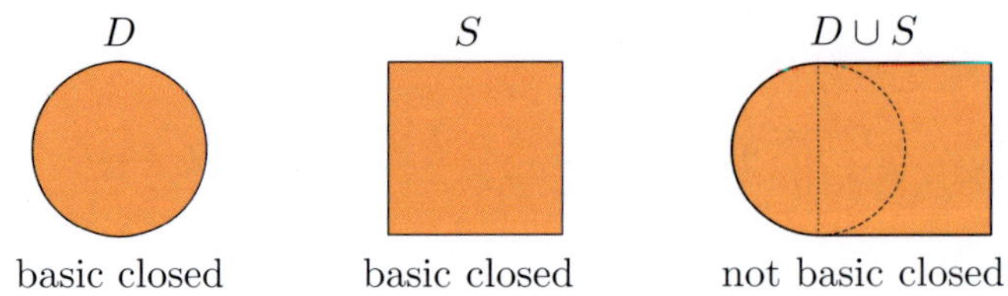

FIGURE 2. The semialgebraic sets of Example 2.2.

To see this, suppose that $D \cup S$ can be described by some set of polynomial inequalities, $g_1(x, y) \geq 0, \ldots, g_s(x, y) \geq 0$. Because this description is finite, there must be some polynomial g_i that changes sign and thus vanishes with odd multiplicity along the left half of the circle defined by $x^2 + y^2 = 1$. That is, for almost every choice of $a \in [-1, 1]$, the univariate polynomial $g_i(x, a) \in \mathbb{R}[x]$ has a root of odd multiplicity at $x = -\sqrt{1-a^2}$. However, any polynomial in $\mathbb{R}[x, y]$ that vanishes on the left half of the circle $\{(-\sqrt{1-a^2}, a) : a \in [-1, 1]\}$ must also vanish on the right half of the circle $\{(\sqrt{1-a^2}, a) : a \in [-1, 1]\}$. In the language of algebraic geometry, the *Zariski closure* of the left half is the entire circle. In particular this is true for g_i and its derivatives $\left(\frac{\partial}{\partial x}\right)^k g_i$ with respect to x. Therefore g_i must also vanish with odd multiplicity on the right half of this circle, implying that g_i is negative at some point in $D \cup S$, and giving a contradiction. $\diamond$

The cone of positive semidefinite matrices is a basic closed semialgebraic set in $\mathcal{S}^N \cong \mathbb{R}^{\binom{N+1}{2}}$ defined by the non-negativity of principal minors. For example,

$$\mathcal{S}_+^2 = \left\{ \begin{pmatrix} a_{11} & a_{12} \\ a_{12} & a_{22} \end{pmatrix} \in \mathcal{S}^2 \ : \ a_{11}a_{22} - a_{12}^2 \geq 0, \ a_{11} \geq 0, \ a_{22} \geq 0 \right\}.$$

More generally, the principal minors $\det(A_{I,I})$ of a symmetric matrix A are polynomials in its entries and $\mathcal{S}_+^N$ is defined by the nonnegativity of all these polynomials. This gives an explicit description of $\mathcal{S}_+^N$ as a basic closed semialgebraic set.

The property of being a basic closed semialgebraic set is inherited by spectrahedra $\mathcal{S}_+^N \cap \mathcal{L}$. (Note that we can impose an equality $\ell(\mathbf{x}) = 0$ by imposing the two inequalities $\ell(\mathbf{x}) \geq 0$ and $-\ell(\mathbf{x}) \geq 0$.) One can also see this in the coordinates $\mathbf{x} = (x_1, \ldots, x_m)$ of the parametrization $\mathcal{L} = \{A(\mathbf{x}) : \mathbf{x} \in \mathbb{R}^m\}$. The spectrahedron $S = \{\mathbf{x} \in \mathbb{R}^m : A(\mathbf{x}) \succeq 0\}$ is defined by the nonnegativity of the principal minors of $A(\mathbf{x})$, which are polynomials in $\mathbf{x}$.

EXAMPLE 1.3 CONT'D. The determinant of the matrix $A(x, y, z)$ in (1.1) is a polynomial of degree four that factors into two quadratics. The semialgebraic description of S given by the principal minors of $A(x, y, z)$ can be simplified to

$$S = \{(x, y, z) \in \mathbb{R}^3 \ : \ (y+x)^2 \leq (z+1)(x+1), (y-x)^2 \leq (z-1)(x-1), x^2 \leq 1\}.$$

Its projection onto coordinates (x, z), seen in Figure 1, is the semialgebraic set

$$\pi(S) = \{(x, z) : -1 \leq z \leq 4x^3 - 3x, x \leq 1/2\} \cup \{(x, z) : 4x^3 - 3x \leq z \leq 1, -1/2 \leq x\}.$$

We can see that $\pi(S)$ is not basic closed using an argument similar to that of Example 2.2. In any polynomial description, $g_1 \geq 0, \ldots, g_s \geq 0$ of $\pi(S)$, some polynomial g_i would have to vanish with odd multiplicity along the curve given by $z = 4x^3 - 3x$, which forms part of the boundary of $\pi(S)$. However this curve

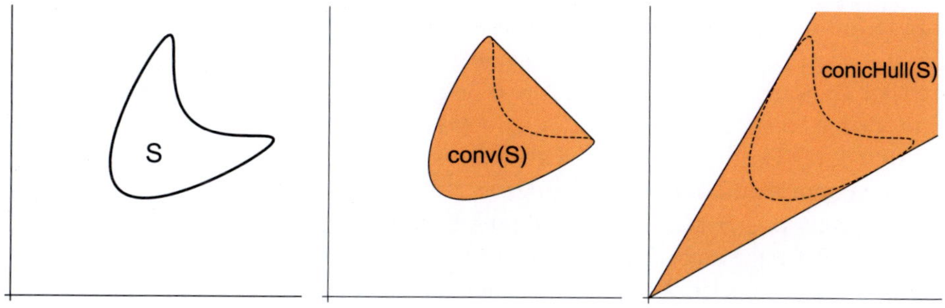

FIGURE 3. The convex and conic hull of a set S in $\mathbb{R}^2$.

also passes through the interior of $\pi(S)$, on which g_i can only vanish with even multiplicity. So no such representation can exist.

Since it is not basic closed, the convex semialgebraic set $\pi(S)$ cannot be written as a spectrahedron! ◇

One fundamental theorem in real algebraic geometry is the Tarski-Seidenberg theorem, which states that the projection of a semialgebraic set is semialgebraic. As we see in Example 1.3, it is not always true that the projection of a basic closed semialgebraic set is basic.

2.2. Convex geometry and faces. Many important features of spectrahedra come from their structure as convex sets. We refer readers to [**1**] for more background on convexity and in particular [**1**, §II.12] for its relevance to the positive semidefinite cone and spectrahedra.

A set $S \subset \mathbb{R}^n$ is *convex* if for any points $\mathbf{x}, \mathbf{y} \in S$, the line segment joining them, $\{\lambda\mathbf{x} + (1-\lambda)\mathbf{y} : 0 \leq \lambda \leq 1\}$ is contained in S. We call S a *convex cone* if it is also invariant by nonnegative scaling, or equivalently if for any two points $\mathbf{x}, \mathbf{y} \in S$, the set of conic combinations $\{\lambda\mathbf{x} + \mu\mathbf{y} : \lambda, \mu \in \mathbb{R}_{\geq 0}\}$ is contained in S.

For example, $\mathcal{S}_+^N$ is a convex cone. To see this, suppose that $A, B \in \mathcal{S}_+^N$. Then for every nonzero vector $\mathbf{v} \in \mathbb{R}^N$, $\mathbf{v}^T A\mathbf{v}$ and $\mathbf{v}^T B\mathbf{v}$ are nonnegative. It follows that for any $\lambda, \mu \in \mathbb{R}_{\geq 0}$, $\mathbf{v}^T(\lambda A + \mu B)\mathbf{v} = \lambda\mathbf{v}^T A\mathbf{v} + \mu\mathbf{v}^T B\mathbf{v} \geq 0$, giving that the conic combination $\lambda A + \mu B$ is also positive semidefinite. One can check that the intersection of any convex sets is convex. Since $\lambda\mathbf{x} + (1-\lambda)\mathbf{y}$ is an affine linear combination of $\mathbf{x}$ and $\mathbf{y}$, any affine linear space is convex. Putting this together with the convexity of $\mathcal{S}_+^N$, we find that a spectrahedron $S = \mathcal{L} \cap \mathcal{S}_+^N$ is convex.

The *convex hull* of a set $S \subset \mathbb{R}^n$ is the smallest convex set containing S and its *conic hull* is the smallest convex cone containing S. See Figure 3. We can write these as

$$\operatorname{conv}(S) = \left\{ \sum_{i=1}^{k} \lambda_i \mathbf{p}_i : \ k \in \mathbb{N}, \ \mathbf{p}_i \in S, \ \lambda_i \geq 0, \ \sum_{i=1}^{k} \lambda_i = 1 \right\}, \text{ and}$$

$$\operatorname{conicHull}(S) = \left\{ \sum_{i=1}^{k} \lambda_i \mathbf{p}_i : \ k \in \mathbb{N}, \ \mathbf{p}_i \in S, \ \lambda_i \geq 0 \right\}.$$

In a convex set some points belong to the convex hull of others. This relationship gives rise to the facial structure of S. A *face* of S is a subset $F \subseteq S$ with the property that for any point in F and every way to write it as a (strict) convex combination of some elements of S, these elements must belong to F. That is, for every $\mathbf{p} \in F$, if $\mathbf{p} = \lambda\mathbf{x} + (1-\lambda)\mathbf{y}$ for some $\lambda \in (0,1)$ where $\mathbf{x}, \mathbf{y} \in S$, then $\mathbf{x}, \mathbf{y} \in F$. Note that the same condition with $\lambda = 0$ or 1 cannot impose any restrictions on the points $\mathbf{x}$ or $\mathbf{y}$, respectively, which is why we restrict to strict convex combinations with $\lambda \in (0,1)$. Both $\emptyset$ and S are always faces of S and we call these the *trivial* faces of S. A nonempty face F of S with $F \neq S$ is called *nontrivial.* An *extreme point* of S is a point $\mathbf{p} \in S$ that forms a singleton face of S. This is a point $\mathbf{p}$ with the property that every way of writing $\mathbf{p}$ as a strict convex combination $\lambda\mathbf{x} + (1-\lambda)\mathbf{y}$ of points $\mathbf{x}$, $\mathbf{y}$ in S with $\lambda \in (0,1)$ satisfies $\mathbf{x} = \mathbf{y} = \mathbf{p}$.

KREIN-MILMAN THEOREM. *A convex compact set in $\mathbb{R}^n$ is the convex hull of its extreme points.*

For any linear function ℓ on $\mathbb{R}^n$, the set of points F maximizing ℓ over S is a face of S. Indeed, if, for some $c \in \mathbb{R}$, the inequality $\ell(\mathbf{x}) \leq c$ holds for all $\mathbf{x} \in S$, then the set of points in S satisfying $\ell(\mathbf{x}) = c$ is a face of S. To see this, note that if $\ell(\mathbf{p}) = c$ and $\mathbf{p} = \lambda\mathbf{x} + (1-\lambda)\mathbf{y}$ for some $\lambda \in (0,1)$ and $\mathbf{x}, \mathbf{y} \in S$, then

$$c = \ell(\mathbf{p}) = \lambda\ell(\mathbf{x}) + (1-\lambda)\ell(\mathbf{y}) \geq \lambda c + (1-\lambda)c = c.$$

Equality of the left and right hand sides forces $\ell(\mathbf{x}) = \ell(\mathbf{y}) = c$.

We call faces of the form $\{\mathbf{x} \in S : \ell(\mathbf{x}) = c\}$ *exposed* and say that ℓ exposes the face. Note that the trivial faces of S are always exposed using the zero function $\ell(\mathbf{x}) = 0$. The set $\emptyset$ is the set of points satisfying $\ell(\mathbf{x}) = 1$ and S is the set of points satisfying $\ell(\mathbf{x}) = 1$. If ℓ is non-zero, then the set of points $\{\mathbf{x} \in \mathbb{R}^n : \ell(\mathbf{x}) = c\}$ forms an affine hyperplane H and $F = S \cap H$. Similarly, if $F = S \cap H$ is a face of S for some affine hyperplane, then it is the set of points maximizing ℓ where $\ell(\mathbf{x}) = c$ is the defining equation for H with signs chosen so that $\ell(\mathbf{x}) \leq c$ for all $\mathbf{x} \in S$. Therefore a nontrivial face of S is exposed if and only if it can be written as $S \cap H$ for some affine hyperplane.

EXAMPLE 2.2 CONT'D. All three sets from Example 2.2 are convex. The nontrivial faces of the disk D are the singleton sets consisting of a point on its boundary, each of which is exposed. For example, the point $(0,1)$ is exposed by the linear function y, as it equals the set of points maximizing y over $(x,y) \in D$. The square S has finitely many faces, namely $\emptyset$, its four vertices, its four edges, and S itself. These faces are all exposed, as highlighted in Figure 4.

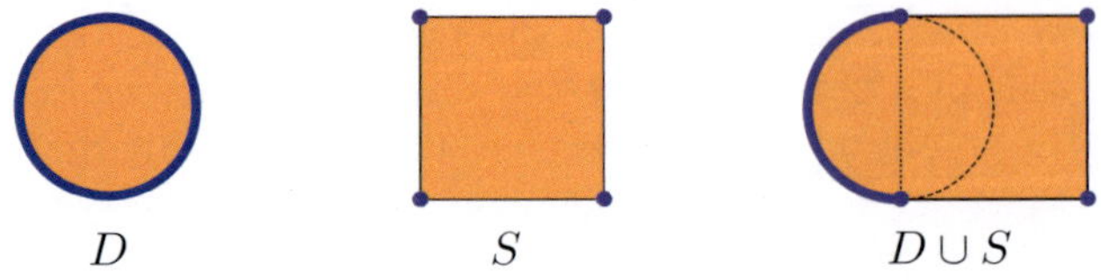

FIGURE 4. Extreme points of the convex sets in Example 2.2.

The facial structure of $D \cup S$ is more complicated. Its nontrivial faces consist of its extreme points, which are the points on the left half of the circle bounding D and two of the vertices of S as well as three edges of S. Note however that the points $(0,1)$ and $(0,-1)$ are not exposed on $D \cup S$. Up to scaling, the only linear function ℓ that maximizes $(0,1)$ over $D \cup S$ is $\ell(x,y) = y$, but every point on the top edge attains this maximum. As promised by the Krein-Milman Theorem, each set is the convex hull of its extreme points. $\diamond$

In their seminal paper on spectrahedra, Ramana and Goldman show that all faces of a spectrahedron are exposed [**13**]. We discuss this theorem further and give a proof in Section 4.2. This gives another way of verifying that a convex semialgebraic set cannot be written as a spectrahedron.

EXAMPLE 1.3 CONT'D. The spectrahedron S in Figure 1 has a curve segment worth of extreme points, namely points of the form $(x, y, z) = (t, 2t^2 - 1, 4t^3 - 3t)$ for $t \in [-1, 1]$. As promised by the Krein-Milman Theorem, S is the convex hull of this curve segment. The other faces of S are the edges joining points on this curve to one of the end points $(\pm 1, 1, \pm 1)$, along with the trivial faces $\emptyset$ and S.

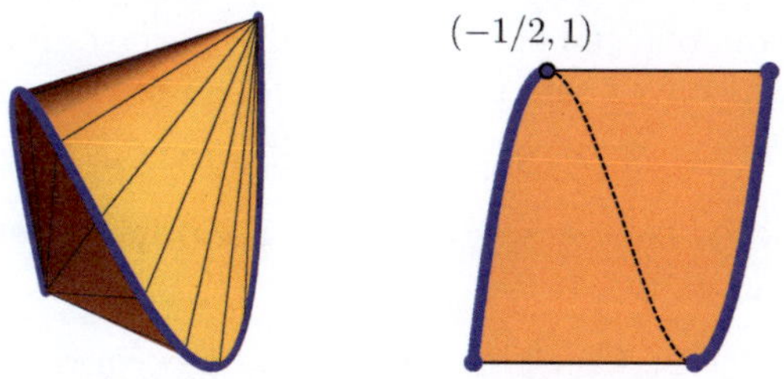

FIGURE 5. Extreme points of the convex sets in Figure 1.

The image $\pi(S)$ under projection $\pi(x, y, z) = (x, z)$ is again a compact, convex set and is the convex hull of its extreme points, shown in Figure 5. The projection $\pi(S)$ also has a non-exposed face. The point $(x, z) = (-1/2, 1)$ is a face of $\pi(S)$, as it cannot be written as a strict convex combination of other points in the set. Up to scaling, the only linear function maximized at this point is $\ell(x, z) = z$, which attains the maximum value $z = 1$. However the face of $\pi(S)$ exposed by this linear function is the line segment $\{(\lambda, 1) : \lambda \in [-1/2, 1]\}$. Therefore the singleton face $\{(-1/2, 1)\}$ of $\pi(S)$ is not exposed. This provides yet another reason that $\pi(S)$ cannot be written as a spectrahedron. $\diamond$

We see here that one way to understand the structure of a convex set is through the affine linear functions that are nonnegative on it. The set of such functions form a convex set in their own right, known as the *dual.*

3. Convex duality

Given a set $S \subset \mathbb{R}^n$, we can define the *dual convex set* of S as

$$S^* = \{\mathbf{c} \in \mathbb{R}^n : 1 + \langle \mathbf{c}, \mathbf{x} \rangle \geq 0 \text{ for all } \mathbf{x} \in S\}.$$

One can check that S^* is a closed convex set that contains the origin. Moreover, if S is semialgebraic, then so is S^*. The definition gives a semialgebraic formula for S^* involving a quantifier. The theory of *quantifier elimination* over $\mathbb{R}$ (see e.g. [**3**]) then implies that S^* has a quantifier-free description as a semialgebraic set.

Some readers may instead be familiar with the *polar* of convex set, given by $S^\circ = \{\mathbf{c} \in \mathbb{R}^n : 1 \geq \langle \mathbf{c}, \mathbf{x} \rangle \text{ for all } \mathbf{x} \in S\}$. But one can check that these two sets are the negative of each other, $S^* = -S^\circ$.

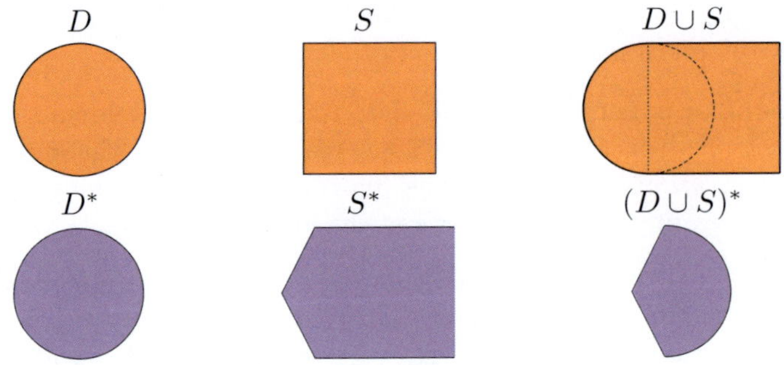

FIGURE 6. The convex sets of Example 2.2 and their duals.

From the definition, one can check that duality reverses inclusions and that the dual of a union of two convex sets is the intersection of their duals, i.e.

$$S \subseteq T \;\Rightarrow\; T^* \subseteq S^* \quad \text{and} \quad (S \cup T)^* \;=\; S^* \cap T^*$$

Moreover, if S is a closed convex set containing the origin, then $(S^*)^* = S$, see e.g. [**1**, Theorem IV.1.2].

EXAMPLE 2.2 CONT'D. Consider the duals of the convex sets in Example 2.2. The dual of each is semialgebraic. In this case, it so happens all three dual sets are basic closed semialgebraic sets, namely

$$\begin{aligned} D^* &= \{(a,b) \in \mathbb{R}^2 : a^2 + b^2 \leq 1\}, \\ S^* &= \{(a,b) \in \mathbb{R}^2 : -1 \leq b \leq 1, -2a-1 \leq b \leq 2a+1\}, \quad \text{and} \\ (D \cup S)^* = D^* \cap S^* &= \{(a,b) \in \mathbb{R}^2 : a^2 + b^2 \leq 1, -2a-1 \leq b \leq 2a+1\}. \end{aligned}$$

These sets are defined by eliminating (x, y) from the inequalities $ax + by + 1 \geq 0$ and $(x, y) \in D$, S, or $D \cup S$, respectively, and pictured in Figure 6.

It is instructive to imagine moving a point (a, b) in the dual set and the corresponding line $ax + by + 1 = 0$ in the plane of the original set, as is done for four points in Figure 7. For any (a, b) in $(D \cup S)^*$, the original, $D \cup S$, belongs to the halfspace $ax+by+1 \geq 0$ bounded by this line, but when (a, b) belong to the *boundary* of the $(D \cup S)^*$, the line intersects $D \cup S$. By considering $((D \cup S)^*)^* = D \cup S$, we can also visualize a point (x, y) in $D \cup S$ as a valid inequality $ax + by + 1 \geq 0$ on points (a, b) in $D \cup S$, as seen in Figure 7. ◇

One convenient interpretation of the dual convex set involves working with convex cones in one dimension higher. Formally, given a set $S \subseteq \mathbb{R}^n$, define the cone over S, denoted cone(S), to be the convex cone in $\mathbb{R}^{n+1}$ defined by

$$\text{cone}(S) \;=\; \{(\lambda, \lambda \mathbf{x}) \;:\; \lambda \in \mathbb{R}_{\geq 0},\ \mathbf{x} \in S\}.$$

Note that we can recover S from its cone by restricting the points with first coordinate equal to one. Furthermore, the operation of coning over sets behaves nicely with duality.

PROPOSITION 3.1. *The operations of taking the cone over a set and taking its dual commute. That is, for a convex set $S \subseteq \mathbb{R}^n$, we have*

$$\text{cone}(S)^* \;=\; \overline{\text{cone}(S^*)}.$$

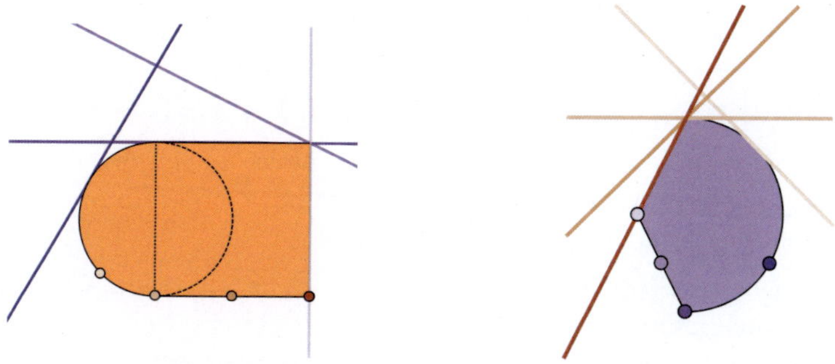

FIGURE 7. Point-line correspondences in a convex set and its dual.

PROOF. Consider a point $(c_0, \mathbf{c})$ in $\mathbb{R}^{n+1}$ with $c_0 > 0$.

$$\begin{aligned}(c_0, \mathbf{c}) \in \text{cone}(S^*) &\Leftrightarrow \tfrac{1}{c_0}\mathbf{c} \in S^* \\ &\Leftrightarrow 1 + \langle \tfrac{1}{c_0}\mathbf{c}, \mathbf{y}\rangle \geq 0 \text{ for all } \mathbf{y} \in S \\ &\Leftrightarrow c_0 + \langle \mathbf{c}, \mathbf{y}\rangle \geq 0 \text{ for all } \mathbf{y} \in S \\ &\Leftrightarrow \lambda(c_0 + \langle \mathbf{c}, \mathbf{y}\rangle) = \langle (c_0, \mathbf{c}), (\lambda, \lambda\mathbf{y})\rangle \geq 0 \text{ for all } \mathbf{y} \in S \text{ and all } \lambda \geq 0 \\ &\Leftrightarrow \langle (c_0, \mathbf{c}), (x_0, \mathbf{x})\rangle \geq 0 \text{ for all } (x_0, \mathbf{x}) \in \text{cone}(S) \\ &\Leftrightarrow (c_0, \mathbf{c}) \in (\text{cone}(S))^*.\end{aligned}$$

Finally we note that both $\text{cone}(S)^*$ and $\text{cone}(S^*)$ belong to the halfspace defined by $c_0 \geq 0$, so to finish, we need only worry about points with $c_0 = 0$.

Note that if $(0, \mathbf{c})$ belongs to $\text{cone}(S)^*$, then $1 + \langle \mathbf{c}, \lambda\mathbf{y}\rangle = \lambda(1/\lambda + \langle \mathbf{c}, \mathbf{y}\rangle) \geq 0$ for all $\mathbf{y} \in S$ and $\lambda \geq 0$. In particular, then $\epsilon + \langle \mathbf{c}, \mathbf{y}\rangle$ is also nonnegative for $\mathbf{y} \in S$ for $\epsilon > 0$. Then $(\epsilon, \mathbf{c})$ belongs to $\text{cone}(S)^*$ for every $\epsilon > 0$ and by the arguments above also $\text{cone}(S^*)$. Taking ϵ to zero then shows that $(0, \mathbf{c})$ belongs to the closure of this cone. Similarly, if $(0, \mathbf{c})$ belongs to the closure of $\text{cone}(S^*)$, then it is a limit of points in this cone with nonzero first coordinate, which must also belong to $\text{cone}(S)^*$. By definition, $\text{cone}(S)^*$ is closed, so it also contains these points. □

In particular, this shows that to understand duality of convex sets (up to issues of closure), it suffices to understand duality of convex cones, which is in many ways simpler. If K is a convex cone, then it is invariant under multiplication by nonnegative scalars. Then $\mathbf{c}$ belongs to K^* if and only if $1 + \langle \mathbf{c}, \lambda\mathbf{x}\rangle = 1 + \langle \lambda\mathbf{c}, \mathbf{x}\rangle$ is nonnegative for all $\mathbf{x} \in K$ and $\lambda \geq 0$. From this, we see that K^* is also a convex cone, known as the *dual cone* of K, and

$$K^* = \{\mathbf{c} \in \mathbb{R}^n : \langle \mathbf{c}, \mathbf{x}\rangle \geq 0 \text{ for all } \mathbf{x} \in K\}.$$

One can check that the following hold for any closed convex cones $C, K \subseteq \mathbb{R}^n$:

$$(C^*)^* = C \quad \text{and} \quad (C \cap K)^* = \overline{C^* + K^*}. \tag{3.1}$$

More formally, given a convex cone K in a real vector space V, its dual is the set of linear functions on V that are nonnegative on K,

$$K^* = \{\ell \in V^* : \ell(\mathbf{x}) \geq 0 \text{ for all } \mathbf{x} \in K\}.$$

Fixing an inner product identifies the vector space V with its dual V^*. For example, under the usual inner product $\langle \mathbf{x}, \mathbf{y} \rangle$ on $\mathbb{R}^n$, the nonnegative orthant is self-dual:

$$(\mathbb{R}^n_{\geq 0})^* \;=\; \{\mathbf{y} \in \mathbb{R}^n : \langle \mathbf{x}, \mathbf{y} \rangle \geq 0 \text{ for all } \mathbf{x} \in \mathbb{R}^n_{\geq 0}\} \;=\; \mathbb{R}^n_{\geq 0}.$$

The self-duality of the nonnegative orthant plays an important role in many algorithms for linear programming. For semidefinite programming, the cone of positive semidefinite matrices plays a similar role.

PROPOSITION 3.2. *Under the inner product* $\langle X, Y \rangle = \text{trace}(XY)$, *the cone of positive semidefinite matrices is self-dual:*

$$(\mathcal{S}^N_+)^* \;=\; \{Y \in \mathcal{S}^N : \langle X, Y \rangle \geq 0 \textit{ for all } X \in \mathcal{S}^N_+\} \;=\; \mathcal{S}^N_+.$$

PROOF. If Y is positive semidefinite, then it can be written as BB^T for some real matrix B. If X is also positive semidefinite, then we can check that the inner product $\langle X, BB^T \rangle = \text{trace}(XBB^T) = \text{trace}(B^T XB)$ is nonnegative, since $B^T XB$ is positive semidefinite. On the other hand, if Y is not positive semidefinite, then there is some vector $\mathbf{v} \in \mathbb{R}^N$ for which $\mathbf{v}^T Y \mathbf{v} < 0$. The rank-one matrix $X = \mathbf{v}\mathbf{v}^T$ belong to $\mathcal{S}^N_+$, but its inner product with Y is negative: $\langle \mathbf{v}\mathbf{v}^T, Y \rangle = \mathbf{v}^T Y \mathbf{v} < 0$. Therefore Y does not belong to the dual cone of $\mathcal{S}^N_+$. □

3.1. Dual operations: linear restriction and projection. Another important property of convex duality is how it interacts with linear projection and restriction. Let L be a m-dimensional linear subspace of $\mathbb{R}^n$. Given a basis $\{\mathbf{a}_1, \ldots, \mathbf{a}_m\}$ for L, consider the maps

$$\pi_L : \mathbb{R}^n \to \mathbb{R}^m \quad \text{given by} \quad \pi_L(\mathbf{v}) = (\langle \mathbf{a}_i, \mathbf{v} \rangle)_{i=1,\ldots,m}, \quad \text{and}$$

$$\pi_L^* : \mathbb{R}^m \to \mathbb{R}^n \quad \text{given by} \quad \pi_L^*(\mathbf{x}) = \sum_{i=1}^m x_i \mathbf{a}_i.$$

Explicitly, if A is an $m \times n$ matrix with rows $\{\mathbf{a}_1, \ldots, \mathbf{a}_m\}$, then these maps are defined by $\pi_L(\mathbf{v}) = A\mathbf{v}$ and $\pi_L^*(\mathbf{x}) = A^T \mathbf{x}$. Note that π_L is surjective and the image is isomorphic to L. Its kernel consists of $L^\perp = \{\mathbf{v} \in \mathbb{R}^n : \langle \mathbf{u}, \mathbf{v} \rangle = 0 \text{ for all } \mathbf{u} \in L\}$. Similarly, the pullback π_L^* is injective and its image equals the linear space L.

Any linear function on $\mathbb{R}^m$ defines a map on $\mathbb{R}^n$ via the pullback π_L^*. Explicitly, given a linear function $\mathbf{y} \mapsto \langle \mathbf{x}, \mathbf{y} \rangle$ on $\mathbb{R}^m$, its pullback is given by $\mathbf{v} \mapsto \langle \mathbf{x}, \pi_L(\mathbf{v}) \rangle$. Then, as the notation suggests, this linear function sends $\mathbf{v} \in \mathbb{R}^n$ to

$$\langle \mathbf{x}, \pi_L(\mathbf{v}) \rangle = \sum_{i=1}^m x_i \langle \mathbf{a}_i, \mathbf{v} \rangle = \left\langle \sum_{i=1}^m x_i \mathbf{a}_i, \mathbf{v} \right\rangle = \langle \pi_L^*(\mathbf{x}), \mathbf{v} \rangle.$$

Under our identification of $(\mathbb{R}^n)^*$ with $\mathbb{R}^n$, this linear function is identified with the point $\pi_L^*(\mathbf{x})$.

Now let $C \subseteq \mathbb{R}^n$ be a convex cone and consider its image, $\pi_L(C)$ in $\mathbb{R}^m$. We would like to relate the dual cone of $\pi_L(C)$ to that of C. A linear function $\langle \mathbf{x}, \mathbf{y} \rangle$ is nonnegative for all $\mathbf{y} \in \pi_L(C)$ if and only if its pullback $\mathbf{v} \mapsto \langle \mathbf{x}, \pi_L(\mathbf{v}) \rangle = \langle \pi_L^*(\mathbf{x}), \mathbf{v} \rangle$ is nonnegative for all $\mathbf{v} \in C$. This happens if and only if $\pi_L^*(\mathbf{x})$ belongs to the dual cone of C. See Figure 8. All together this gives that

$$(\pi_L(C))^* = \{\mathbf{x} \in \mathbb{R}^m : \pi_L^*(\mathbf{x}) \in C^*\} \;\cong\; C^* \cap L. \tag{3.2}$$

The last equality here uses that π_L^* is a linear isomorphism between $\mathbb{R}^m$ and L. This shows that intersection with a linear space is the dual operation to linear projection.

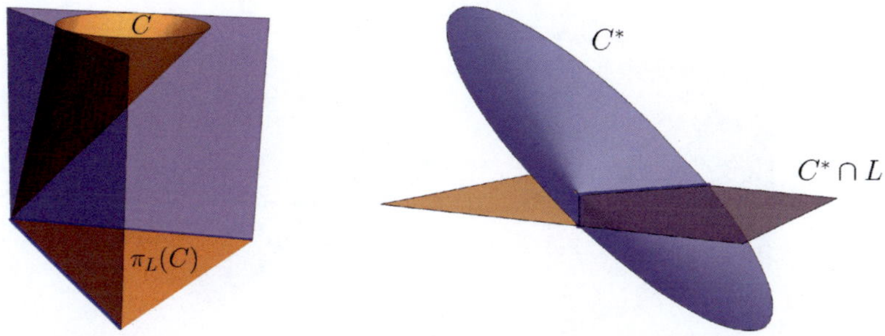

FIGURE 8. Nonnegative linear functions on a projection $\pi_L(C)$ come from nonnegative linear functions C that are constant along preimages. These constitute a linear slice of the dual cone C^*.

Recall that dualizing any convex cone K twice results in its closure, $(K^*)^* = \overline{K}$. Applying the dual to both sides of the equation above gives that

$$\overline{\pi_L(C)} = (\{\mathbf{x} \in \mathbb{R}^m : \pi_L^*(\mathbf{x}) \in C^*\})^* \quad \cong \quad (C^* \cap L)^*.$$

One way to interpret this statement is that the dual cone of a linear slice $C^* \cap L$ is isomorphic to the dual cone $C = (C^*)^*$ under projection. All together this gives the following:

PROPOSITION 3.3. *The dual cone of the image of C under linear map is linearly isomorphic to a linear section of C^*, namely*

$$(\pi_L(C))^* \quad \cong \quad C^* \cap L.$$

The dual cone of a linear section of C is the image of C^* under a linear map:

$$(C \cap L)^* \quad \cong \quad \overline{\pi_L(C^*)}.$$

Here $(C \cap L)^*$ is the convex dual of $(C \cap L)^*$ in the ambient space L. In short, projection and slicing are dual operations.

For the study of spectrahedra, an important special case is the cone of positive semidefinite matrices $C = C^* = \mathcal{S}_+^N$. If $\mathcal{L}$ is a linear space spanned by matrices $A_1, \ldots, A_m \in \mathcal{S}^N$, then we can consider the linear map

$$\pi_{\mathcal{L}} : \mathcal{S}^N \to \mathbb{R}^m \quad \text{given by } X \mapsto (\langle A_1, X\rangle, \ldots, \langle A_m, X\rangle).$$

A function $\ell(\mathbf{y}) = \langle \mathbf{x}, \mathbf{y}\rangle$ represented by $\mathbf{x} \in \mathbb{R}^m$ is nonnegative on $\pi_{\mathcal{L}}(\mathcal{S}_+^N)$ if and only if the function on $\mathcal{S}^N$ given by $X \mapsto \langle \pi_{\mathcal{L}}^*(\mathbf{x}), X\rangle = \langle \sum_{i=1}^m x_i A_i, X\rangle$ is nonnegative on $\mathcal{S}_+^N$. This happens exactly when the matrix $\sum_{i=1}^m x_i A_i$ is positive semidefinite. Therefore the dual cone of the projection is given by

$$\pi_{\mathcal{L}}(\mathcal{S}_+^N)^* \quad = \quad \left\{ \mathbf{x} \in \mathbb{R}^m \quad : \quad \sum_{i=1}^m x_i A_i \succeq 0 \right\} \quad \cong \quad \mathcal{S}_+^N \cap \mathcal{L}. \tag{3.3}$$

The image of $\mathcal{S}_+^N$ under the projection $\pi_{\mathcal{L}}$ may not be closed. For example, consider the image of $\mathcal{S}_+^2$ under the map $X \mapsto (X_{11}, X_{12})$. This contains points of the form $(X_{11}, X_{12}) = (\epsilon, 1)$ for every $\epsilon > 0$, coming from matrices with $X_{22} > 1/\epsilon$. However the point $(0, 1)$ does not belong to this image, as there is no positive semidefinite matrix X with $X_{11} = 0$ and $X_{12} > 0$.

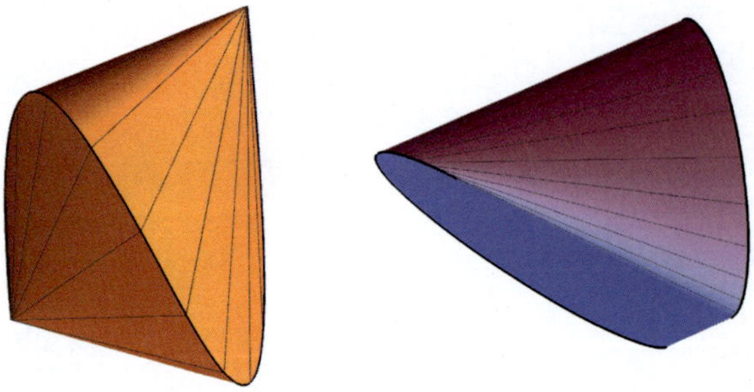

FIGURE 9. A spectrahedron and its dual convex body

However for any convex (not necessarily closed) cone $K \subset \mathbb{R}^n$, the dual of its dual equals its closure, i.e. $(K^*)^* = \overline{K}$. Taking the dual of the cones in (3.3), we find that the dual of a spectrahedral cone is the image of $\mathcal{S}_+^N$ under a linear map:

$$\overline{\pi_{\mathcal{L}}(\mathcal{S}_+^N)} \;=\; \left(\left\{\mathbf{x} \in \mathbb{R}^m \;:\; \sum_{i=1}^m x_i A_i \succeq 0\right\}\right)^* \;\cong\; (\mathcal{S}_+^N \cap \mathcal{L})^*. \tag{3.4}$$

It is also instructive to apply Proposition 3.3 to spectrahedral cones C and their images under linear maps. In particular, unlike spectrahedra, the class of spectrahedral shadows is closed under duality.

PROPOSITION 3.4. *The dual of a spectrahedral shadow is (the closure of) a spectrahedral shadow.*

PROOF. By Proposition 3.1, it suffices to work with a spectrahedral cone, i.e. the intersection of $\mathcal{S}_+^N$ with a linear space (rather than an affine linear space).

Consider the affine space $\mathcal{L}$ spanned by matrices $A_1, \ldots, A_n$ and the spectrahedral cone $C = \{\mathbf{x} \in \mathbb{R}^n \text{ s.t. } \sum_{i=1}^n A_i \succeq 0\} \cong \mathcal{S}_+^N \cap \mathcal{L}$. For $m < n$, we can examine the image of C under the projection $\pi_m : \mathbb{R}^n \to \mathbb{R}^m$ given by $\pi_m(\mathbf{x}) = (x_1, \ldots, x_m)$. The pullback $\pi_m^* : \mathbb{R}^m \to \mathbb{R}^n$ is given by $\pi_m^*(\mathbf{y}) = (\mathbf{y}, \mathbf{0})$. Then by (3.2),

$$(\pi_m(C))^* = \{\mathbf{y} \in \mathbb{R}^m : (\mathbf{y}, \mathbf{0}) \in C^*\}.$$

The dual cone C^* is the projection of the cone of positive semidefinite matrices: $C^* = \overline{\pi_{\mathcal{L}}(\mathcal{S}_+^N)} = \overline{\{(\langle X, A_i\rangle)_{i=1,\ldots n} : X \in \mathcal{S}_+^N\}}$. Putting this together, we find that

$$(\pi_m(C))^* = \overline{\{(\langle X, A_i\rangle)_{i=1,\ldots m} : X \in \mathcal{S}_+^N, \langle X, A_j\rangle = 0 \text{ for } j = m+1, \ldots, n\}}.$$

This is the image of the spectrahedron $\{X \in \mathcal{S}_+^N : \langle X, A_j\rangle = 0 \text{ for } j = m+1, \ldots, n\}$ under the linear map $X \mapsto (\langle X, A_i\rangle)_{i=1,\ldots m}$ up to closure. □

EXAMPLE 1.3 CONT'D. Let $A_0, A_1, A_2, A_3 \in \mathcal{S}^4$ be matrices with $(A_k)_{ij} = 1$ if $|i-j| = k$ and 0 otherwise for every $k = 0, 1, 2, 3$. The cone over the spectrahedron $C = \text{cone}(S)$ from Example 1.3 can be written as the intersection of the positive semidefinite cone with the linear space spanned by these matrices

$$C \;=\; \{(w, x, y, z) \in \mathbb{R}^4 : wA_0 + xA_1 + yA_2 + zA_3 \succeq 0\} \;\cong\; \mathcal{S}_+^4 \cap \mathcal{L}$$

where $\mathcal{L} = \text{span}_{\mathbb{R}}\{A_0, A_1, A_2, A_3\}$. Since the cone $\mathcal{S}_+^4$ is self-dual, the dual cone of C is given by the projection of $\mathcal{S}_+^4$ onto this linear space:

$$C^* \;=\; \{(\langle A_0, X\rangle, \langle A_1, X\rangle, \langle A_2, X\rangle, \langle A_3, X\rangle) \in \mathbb{R}^4 : X \in \mathcal{S}_+^4\} \;\cong\; \pi_{\mathcal{L}}(\mathcal{S}_+^4).$$

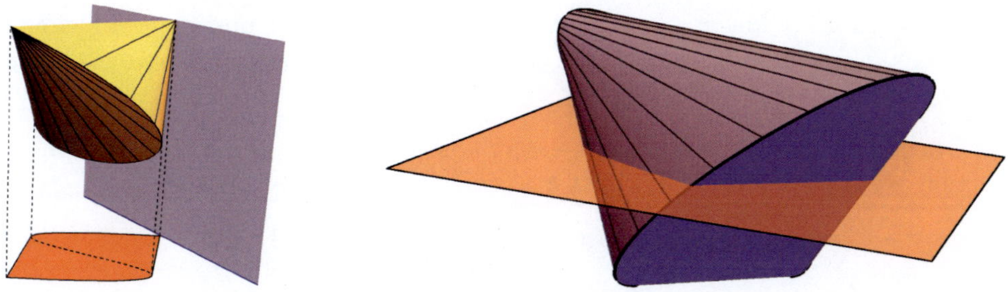

FIGURE 10. A projection of the spectrahedron from Example 1.3 and corresponding slice of its dual convex body.

To recover the original set S and its convex dual set S^*, we can restrict both sets to have first coordinate one, giving the convex sets

$$S = \{(x,y,z) \in \mathbb{R}^3 : A_0 + xA_1 + yA_2 + zA_3 \succeq 0\} \text{ and}$$
$$S^* = \{(\langle A_1, X\rangle, \langle A_2, X\rangle, \langle A_3, X\rangle) \in \mathbb{R}^3 : X \in \mathcal{S}^4_+, \langle A_0, X\rangle = 1\},$$

seen in Figure 9. Taking things one step further, we can consider the projection of S on to the coordinates (x,z),

$$\pi(S) = \{(x,z) \in \mathbb{R}^2 : \exists\, y \in \mathbb{R} \text{ with } A_0 + xA_1 + yA_2 + zA_3 \succeq 0\}$$

as shown in Figure 10.

To understand the dual convex body, we consider affine linear functions on $\mathbb{R}^3$ that come from affine linear functions on $\mathbb{R}^2$. These are exactly the functions $(x,y,z) \mapsto 1 + ax + by + cz$ where $(a,b,c) \in \mathbb{R}^3$ and $b = 0$. This lets us write the dual convex body as the intersection of S^* with the plane $b = 0$, giving

$$(\pi(S))^* = \{(\langle A_1, X\rangle, \langle A_3, X\rangle) \in \mathbb{R}^3 : X \in \mathcal{S}^4_+, \langle A_0, X\rangle = 1, \langle A_2, X\rangle = 0\}.$$

This is also the image of a spectrahedron under a linear map, namely the image of the spectrahedron $\{X \in \mathcal{S}^4_+ : \langle A_0, X\rangle = 1, \langle A_2, X\rangle = 0\}$ under the linear map $X \mapsto (\langle A_1, X\rangle, \langle A_3, X\rangle)$. Indeed, the self-duality of the cone of positive semidefinite matrices means that the class of projected spectrahedra is closed under duality. $\diamond$

One consequence of this section is that any linear function on the projection of a spectrahedron can be lifted to a linear function on the original spectrahedron. Therefore optimizing a linear function over a spectrahedral shadow can also be phrased as a semidefinite program, which we discuss in Section 3.2.

3.2. Convex optimization and semidefinite programming. A *conic optimization problem* has the form

$$\min_{\mathbf{x} \in K} \langle \mathbf{c}, \mathbf{x}\rangle \quad \text{such that} \quad \langle \mathbf{a}_i, \mathbf{x}\rangle = b_i \quad \text{for} \quad i = 1, \ldots, m,$$

where K is a convex cone in $\mathbb{R}^n$, $\mathbf{c}, \mathbf{a}_1, \ldots, \mathbf{a}_m \in \mathbb{R}^n$ and $b_1, \ldots, b_m \in \mathbb{R}$. We call a point $\mathbf{x}$ *feasible* if it satisfies the constraints $\mathbf{x} \in K$ and $\langle \mathbf{a}_i, \mathbf{x}\rangle = b_i$ for all i. Many optimization problems can be reformulated as conic optimization problems. One benefit of this approach is that most of the complexity of the problem is packaged into the convex cone, rather than both the feasible region and objective function.

When the cone K is the nonnegative orthant, the feasible region $\{\mathbf{x} \in \mathbb{R}^n_{\geq 0} : \langle \mathbf{a}_i, \mathbf{x}\rangle = b_i \text{ for } i = 1, \ldots, m\}$, is a *polyhedron* and corresponding conic optimization problem is called a *linear program*. When K is the cone of positive semidefinite matrices, the feasible region is a *spectrahedron* and the convex program is called a *semidefinite program*.

DEFINITION 3.5. A *semidefinite program* is the problem of minimizing a linear function over a spectrahedron, that is, a problem of the form

$$\min_{X \in \mathcal{S}_+^N} \langle C, X\rangle \quad \text{subject to } \langle A_i, X\rangle = b_i \text{ for } i = 1, \dots, m,$$

where $C, A_1, \dots, A_m \in \mathcal{S}^N$ and $b_1, \dots, b_m \in \mathbb{R}$.

The feasible region, $\{X \in \mathcal{S}_+^N : \langle A_i, X\rangle = b_i \text{ for } i = 1, \dots, m\}$, of a semidefinite program is a spectrahedron, as it is the intersection of the positive semidefinite cone with the affine linear space of matrices satisfying the constraints.

Semidefinite programs are useful because of their ability to model a large range of problems and because they can be solved efficiently using interior point methods. See [**18**] for more details on algorithms and applications.

EXAMPLE 1.3 CONT'D. Let E_{ij} denote the 4×4 real symmetric matrix with (i, j)th and (j, i)th entry 1 and all other entries 0. The spectrahedron S given by the set of matrices $X \in \mathcal{S}_+^4$ satisfying

$$\langle E_{ii}, X\rangle = 1 \text{ for all } i \text{ and } \langle E_{12} - E_{23}, X\rangle = \langle E_{23} - E_{34}, X\rangle = \langle E_{13} - E_{24}, X\rangle = 0$$

is exactly the spectrahedron given in parametrized form in Example 1.3. Consider the matrix $C = 3E_{12} - E_{13}$. Minimizing the linear function $\langle C, X\rangle$ over S is equivalent to minimizing the linear function $6x - 2z$ over

$$\left\{ (x, y, z) \in \mathbb{R}^3 : \begin{pmatrix} 1 & x & y & z \\ x & 1 & x & y \\ y & x & 1 & x \\ z & y & x & 1 \end{pmatrix} \succeq 0 \right\}.$$

The minimum equals $-4\sqrt{2}$, which is achieved by the point $(x, y, z) = (-\frac{1}{\sqrt{2}}, 0, \frac{1}{\sqrt{2}})$, at which point the matrix above is positive semidefinite and has rank two. $\diamond$

More generally, the optimal values of semidefinite programs are semialgebraic functions of the inputs, whose degrees are studied in [**11, 14**].

Another fundamental benefit of conic programming is the ability to easily formulate a "dual problem" that provides lower bounds on the optimal value of the original problem (often called the *primal* problem):

$$\text{(Primal)} \qquad \min_{\mathbf{x} \in K} \langle \mathbf{c}, \mathbf{x}\rangle \qquad \text{s.t.} \qquad \langle \mathbf{a}_i, \mathbf{x}\rangle = b_i \ \text{ for } \ i = 1, \dots, m$$

$$\text{(Dual)} \qquad \max_{\mathbf{y} \in \mathbb{R}^m} \sum_{i=1}^m b_i y_i \qquad \text{s.t.} \qquad \mathbf{c} - \sum_{i=1}^m \mathbf{a}_i y_i \ \in \ K^*$$

Here K^* denotes the dual cone of K.

Note that $\{\mathbf{c} - \sum_{i=1}^m \mathbf{a}_i y_i : \mathbf{y} \in \mathbb{R}^m\}$ parametrizes an affine linear space. So the dual problem still has form of optimizing a linear function over the intersection of a convex cone and an affine space.

THEOREM (Weak duality and complementary slackness). For any primal feasible $\mathbf{x}$ and dual feasible $\mathbf{y}$,

$$\langle \mathbf{c}, \mathbf{x} \rangle \;\geq\; \sum_{i=1}^{m} b_i y_i.$$

Moreover, if $\langle \mathbf{c}, \hat{\mathbf{x}} \rangle = \sum_{i=1}^{m} b_i \hat{y}_i$ for feasible points $\hat{\mathbf{x}}$ and $\hat{\mathbf{y}}$ then both are optimal.

PROOF. Suppose $\mathbf{x}, \mathbf{y}$ are feasible. Then

$$\begin{aligned} \langle \mathbf{c}, \mathbf{x} \rangle - \sum_{i=1}^{m} b_i y_i &= \langle \mathbf{c}, \mathbf{x} \rangle - \sum_{i=1}^{m} \langle \mathbf{a}_i, \mathbf{x} \rangle y_i \\ &= \langle \mathbf{c} - \sum_{i=1}^{m} y_i \mathbf{a}_i, \mathbf{x} \rangle \;\geq\; 0. \end{aligned}$$

Here the last inequality follows from the fact that $\mathbf{c} - \sum_{i=1}^{m} y_i \mathbf{a}_i \in K^*$ and $\mathbf{x} \in K$.

If $\langle \mathbf{c}, \hat{\mathbf{x}} \rangle = \sum_{i=1}^{m} b_i \hat{y}_i$, then $\langle \mathbf{c}, \hat{\mathbf{x}} \rangle$ is an upper bound for $\sum_i b_i y_i$ over all feasible $\mathbf{y}$. Since $\hat{\mathbf{y}}$ achieves this upper bound, it must be optimal. Similarly, $\sum_i b_i \hat{y}_i$ is a lower bound for $\langle \mathbf{c}, \mathbf{x} \rangle$, which is achieved by $\hat{\mathbf{x}}$. □

4. Facial structure, extreme points and ranks

In this section, we discuss in detail the facial structure and extreme points of spectrahedra. We also explore this theory in more detail for the Gram spectrahedron associated to a nonnegative univariate polynomial. To do this, we first need to understand the facial structure of $\mathcal{S}_+^N$.

4.1. Faces of the PSD cone. The facial structure of $\mathcal{S}_+^N$ is governed by linear subspaces parametrizing their kernels. This is discussed at length in [**1**, §II.12] and we go over some of the relevant proofs. In particular, we show the following bijection between linear subspaces of $\mathbb{R}^N$ and faces of the convex cone $\mathcal{S}_+^N$.

THEOREM 4.1. *For every linear subspace $L \subseteq \mathbb{R}^N$,*

$$\mathcal{F}_L \;=\; \left\{ A \in \mathcal{S}_+^N \;:\; L \subseteq \ker(A) \right\}$$

is a face of the cone $\mathcal{S}_+^N$. Moreover every face $\mathcal{S}_+^N$ has this form.

To prove this theorem, we need the following fundamental facts about positive semidefinite matrices.

LEMMA 4.2. *For $A, B \in \mathcal{S}_+^N$ and $\mathbf{v} \in \mathbb{R}^N$,*

(i) $\mathbf{v}^T A \mathbf{v} = 0$ *if and only if* $A\mathbf{v} = 0$,
(ii) $\langle A, B \rangle = 0$ *if and only if* $\operatorname{rowspan}(B) \subseteq \ker(A)$, *and*
(iii) $\ker(A + B) = \ker(A) \cap \ker(B)$.

PROOF. For (i), note that we can write the matrix A as a sum of rank-one matrices, $A = \sum_{i=1}^{d} \mathbf{u}_i \mathbf{u}_i^T$. Then

$$\mathbf{v}^T A \mathbf{v} \;=\; \sum_{j=1}^{d} \mathbf{v}^T \mathbf{u}_i \mathbf{u}_i^T \mathbf{v} \;=\; \sum_{i=1}^{d} (\langle \mathbf{u}_i, \mathbf{v} \rangle)^2.$$

As a sum of nonnegative terms, this equals zero if and only if each term $(\langle \mathbf{u}_i, \mathbf{v} \rangle)^2$ is zero. In particular, if $\mathbf{v}^T A \mathbf{v}$ is zero, then $\langle \mathbf{u}_i, \mathbf{v} \rangle = \mathbf{u}_i^T \mathbf{v}$ is zero for each i, implying that $A\mathbf{v} = \sum_i \mathbf{u}_i \mathbf{u}_i^T \mathbf{v}$ is zero. The other direction is clear.

(ii) Similarly if B is positive semidefinite, we can write $B = \sum_{j=1}^{e} \mathbf{v}_j\mathbf{v}_j^T$. Then

$$\langle A, B\rangle \;=\; \sum_{i=1}^{e}\langle A, \mathbf{v}_j\mathbf{v}_j^T\rangle \;=\; \sum_{j=1}^{e}\mathbf{v}_j^T A\mathbf{v}_j.$$

Since A is positive semidefinite, each term $\mathbf{v}_j^T A\mathbf{v}_j$ is nonnegative. The sum is zero if and only if each summand is. By part (i), this occurs if and only if $\mathbf{v}_j \in \ker(A)$ for all j. This shows that the rowspan of B is contained in the kernel of A.

Finally, for (iii), note that any vector in the kernel of both A and B is automatically contained in the kernel of their sum. For the reverse inclusion, suppose that $\mathbf{v}$ belongs to $\ker(A+B)$. In particular,

$$\mathbf{v}^T(A+B)\mathbf{v} \;=\; \mathbf{v}^T A\mathbf{v} + \mathbf{v}^T B\mathbf{v} \;=\; 0.$$

Since $A, B \in \mathcal{S}_+^N$, both summands are nonnegative, implying that they are both zero. Then by (i), $\mathbf{v}$ belongs to the kernel of both A and B. $\square$

Now we are ready to prove the theorem characterizing faces of $\mathcal{S}_+^N$.

Proof of Theorem 4.1. For the first statement, suppose that $A, B \in \mathcal{S}_+^N$ and $\lambda A + \mu B \in \mathcal{F}_L$ for some $\lambda, \mu > 0$. Then

$$L \;\subseteq\; \ker(\lambda A + \mu B) \;=\; \ker(\lambda A)\cap\ker(\mu B) \;=\; \ker(A)\cap\ker(B).$$

The subspace L belongs to the kernels of both A and B, meaning that $A, B \in \mathcal{F}_L$. This shows that $\mathcal{F}_L$ is a face of $\mathcal{S}_+^N$.

Now suppose that F is any face of $\mathcal{S}_+^N$. Let $X \in F$ be a matrix of maximal rank in F and let $L = \ker(X)$. First we claim that $F \subseteq \mathcal{F}_L$. To see this, let A be any element of F. Since F is a convex cone, $A + X$ also belongs to F and so, by assumption, the rank of $A + X$ cannot be larger than that of X. In particular, the dimension of the kernel of $A + X$ cannot be smaller than the dimension of $\ker(X)$. Lemma 4.2 then implies that $\ker(A) \cap \ker(X) = \ker(X)$, meaning that the kernel of A contains $L = \ker(X)$.

To show the reverse inclusion $\mathcal{F}_L \subseteq F$, let $B \in \mathcal{F}_L$. We claim that for sufficiently large $\lambda \in \mathbb{R}_+$, the matrix $A = \lambda X - B$ is positive semidefinite. To see this, consider the linear space $L^\perp$ consisting of vectors $\mathbf{v}$ with $\langle \mathbf{u}, \mathbf{v}\rangle = 0$ for all $\mathbf{u} \in L$ and the compact set

$$S = \left\{\mathbf{v} \in L^\perp : ||\mathbf{v}||_2 = 1\right\}.$$

Since X is positive semidefinite, the function $\mathbf{v} \mapsto \mathbf{v}^T X\mathbf{v}$ is nonnegative on S. Moreover, its minimum over X must be strictly positive, since no vector in S can belong to $L = \ker(X)$. Let m be this minimum. Similarly, the function $\mathbf{v} \mapsto \mathbf{v}^T B\mathbf{v}$ is nonnegative and attains some maximum M on S. For $\lambda > M/m$, the function $\mathbf{v} \mapsto \mathbf{v}^T(\lambda X - B)\mathbf{v}$ is positive on S. Indeed we check that

$$\mathbf{v}^T(\lambda X - B)\mathbf{v} = \lambda(\mathbf{v}^T X\mathbf{v}) - (\mathbf{v}^T B\mathbf{v}) \geq \lambda m - M > 0.$$

By scaling, we see that this function is nonnegative for all $\mathbf{v} \in L^\perp$. Note that any vector $\mathbf{x} \in \mathbb{R}^N$ can be written as $\mathbf{u} + \mathbf{v}$ where $\mathbf{u} \in L$ and $\mathbf{v} \in L^\perp$. If $A = \lambda X - B$, then, by assumption, $\mathbf{u}$ belongs to the kernel of A. This gives that

$$\mathbf{x}^T A\mathbf{x} \;=\; \mathbf{u}^T A\mathbf{u} + 2\mathbf{v}^T A\mathbf{u} + \mathbf{v}^T A\mathbf{v} \;=\; \mathbf{v}^T A\mathbf{v} \geq 0.$$

Now we have $A, B \in \mathcal{S}_+^N$ with $A + B = \lambda X \in F$. Since F is a face of $\mathcal{S}_+^N$, it follows that both A and B belong to F. This shows $\mathcal{F}_L \subseteq F$, as desired. $\square$

COROLLARY 4.3. Every face of $\mathcal{S}_+^N$ is exposed.

PROOF. By the above theorem, it suffices to show that for any subspace $L \subset \mathbb{R}^N$, the face $\mathcal{F}_L$ is exposed. Let B be a positive semidefinite matrix whose rowspan equals L. To construct such a matrix, we can, for example, take a set of vectors $\mathbf{v}_1, \ldots, \mathbf{v}_d$ spanning L and consider the positive semidefinite matrix $B = \sum_{i=1}^d \mathbf{v}_i \mathbf{v}_i^T$. By Lemma 4.2, a matrix $A \in \mathcal{S}_+^N$ satisfies $\langle A, B\rangle = 0$ if and only if $\text{rowspan}(B) = L \subseteq \ker(A)$. This shows that the face $\mathcal{F}_L$ is exposed by the linear function $\langle \cdot, B\rangle$. □

This theorem shows that every face of $\mathcal{S}_+^N$ is linearly isomorphic to $\mathcal{S}_+^k$ for some $k \leq N$. Given a linear space $L \subseteq \mathbb{R}^N$ of dimension d, consider an invertible $N \times N$ matrix U whose last d columns span L. That is, for any vector $\mathbf{x}$ of the form $(\mathbf{0}, \mathbf{v})$ with $\mathbf{v} \in \mathbb{R}^d$, $U\mathbf{v} \in L$. Then $X \mapsto U^T X U$ is an invertible linear transformation on $\mathcal{S}^N$ with the property that

$$A \in \mathcal{F}_L \quad \Leftrightarrow \quad U^T A U = \begin{pmatrix} \tilde{A} & 0 \\ 0 & 0 \end{pmatrix} \text{ for some } \tilde{A} \in \mathcal{S}_+^{N-d}.$$

This shows that $\mathcal{F}_L$ is linearly isomorphic to the cone $\mathcal{S}_+^{N-d}$.

4.2. Faces of spectrahedra. Many nice properties of $\mathcal{S}_+^N$ are inherited by spectrahedra. The first of which is that faces are characterized by kernels.

THEOREM 4.4. *Every face of a spectrahedron $S \subset \mathcal{S}_+^N$ has the form*

$$F \;=\; \{A \in S \;:\; L \subseteq \ker(A)\}$$

for some linear space $L \subseteq \mathbb{R}^N$. That is, $F = S \cap \mathcal{F}_L$, as in Theorem 4.1.

Here we need to refer to the *relative interior* of a convex set. Given a convex set $S \subset \mathbb{R}^n$, the *relative interior* of S, denoted $\text{ri}(S)$, is the interior of S within its affine span. By [**4**, Theorem I.5.6], given a closed convex set S and a face F containing a point $\mathbf{x}$, F is the smallest face of S containing $\mathbf{x}$ if and only if $\mathbf{x}$ belongs to the relative interior of F.

LEMMA 4.5. *Given a closed convex set S and an affine linear space L in $\mathbb{R}^n$, every face F of $S \cap L$ has the form $\tilde{F} \cap L$ for some face $\tilde{F}$ of S. Moreover if $\tilde{F}$ is an exposed face of S then $\tilde{F} \cap L$ is an exposed face of $S \cap L$.*

PROOF. One can check that for every face $\tilde{F}$ of S, the intersection $\tilde{F} \cap L$ is a face of $S \cap L$ and we leave this to the reader.

Let $\mathbf{x}$ be a point in the relative interior of F and let $\tilde{F}$ be the smallest face of S containing $\mathbf{x}$. Then by [**4**, Theorem I.5.6], $\mathbf{x}$ belongs to the relative interior of F. We claim that $\mathbf{x}$ then belongs to the relative interior of $\tilde{F} \cap L$. Let $\mathbf{v}$ be a vector so that $\mathbf{x} + \lambda\mathbf{v}$ belongs to the affine span of $\tilde{F} \cap L$ for all $\lambda \in \mathbb{R}$. It suffices to show that for sufficiently small $\epsilon > 0$, $\mathbf{x} + \epsilon\mathbf{v}$ belongs to $\tilde{F} \cap L$. By assumption, it belongs to L. Moreover, the affine span of $\tilde{F} \cap L$ is contained in that of $\tilde{F}$, so $\mathbf{x} + \lambda\mathbf{v}$ belongs to the affine span of $\tilde{F}$ for all $\lambda \in \mathbb{R}$. Since $\mathbf{x}$ belongs to the relative interior of $\tilde{F}$, there is some $\epsilon > 0$ so that $\mathbf{x} + \epsilon\mathbf{v} \in \tilde{F}$, which proves the claim.

Since $\mathbf{x}$ belongs to both the relative interiors of the faces F and $\tilde{F} \cap L$ of $S \cap L$, they are both the smallest face of S containing $\mathbf{x}$ and therefore must be equal. Finally, if $\ell : \mathbb{R}^n \to \mathbb{R}$ is a linear function exposing the face $\tilde{F}$ of S, then the same linear function exposes the face F of $S \cap L$. □

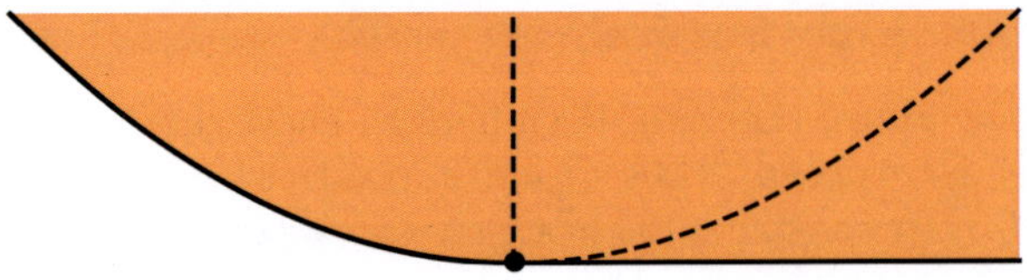

FIGURE 11. A non-exposed face of $\{(a,b) \in \mathbb{R}^2 : t^4 + at^2 + b \in P_{1,\leq 4}\}$.

PROOF OF THEOREM 4.4. A spectrahedron has the form $S = \mathcal{L} \cap \mathcal{S}_+^N$ for some affine linear space $\mathcal{L} \subset \mathcal{S}^N$. By Lemma 4.5, any face of S is the intersection of a face of $\mathcal{S}_+^N$ with $\mathcal{L}$. The characterization in Theorem 4.1 then completes the proof. □

An immediate corollary of this is the theorem of Ramana and Goldman [**13**] that spectrahedra are facially exposed:

COROLLARY 4.6. Every face of a spectrahedron is exposed.

One interesting application is that the cone of nonnegative univariate polynomials of degree ≤ 4 is not a spectrahedron.

EXAMPLE 4.7. See also [**2**, Example 3.13]. Consider the convex cone $P_{1,\leq 4}$ of polynomials $f \in \mathbb{R}[t]$ of degree ≤ 4 that are globally nonnegative. We can show that $P_{1,\leq 4}$ is not a spectrahedron by exhibiting an affine linear section of it with a non-exposed face. Indeed, consider the affine linear space L of polynomials of $\mathbb{R}[t]_{\leq 4} \cong \mathbb{R}^5$ that are monic of degree 4, with no odd degree terms. That is $L = \{t^4 + at^2 + b : a, b \in \mathbb{R}\}$. Note that $f(t) = t^4 + at^2 + b$ is nonnnegative for all $t \in \mathbb{R}$ if and only if the quadratic polynomial $p(t) = t^2 + at + b$ is nonnegative on $\mathbb{R}_{\geq 0}$. This happens if and only if $a, b \geq 0$ or $a^2 \leq 4b$. The point $(a,b) = (0,0)$ is a non-exposed face of the set $\{(a,b) : a, b \geq 0 \text{ or } a^2 \leq 4b\} \cong L \cap P_{1,\leq 4}$, seen in Figure 11. By Lemma 4.5, the polynomial $f = t^4$ corresponding to this point belongs to a non-exposed face of the convex cone $P_{1,\leq 4}$. Therefore $P_{1,\leq 4}$ cannot be written as a spectrahedron.

4.3. The Pataki range. The minimum of a linear function over a compact, convex set is always achieved at an extreme point. When this set is a spectrahedron, then the extreme point will be associated to a matrix of some rank. Such a point belongs to the boundary, and so this matrix necessarily drops rank. In fact it must satisfy much stronger inequalities. For spectrahedra obtained from generic affine linear spaces, the ranks must belong to what is often called the *Pataki range*, due to its introduction by Gabor Pataki in [**12**].

THEOREM 4.8 (Pataki range). *Let $A_0, \ldots, A_m \in \mathcal{S}^N$ be linearly independent over $\mathbb{R}$ and consider the spectrahedron $S = \{\mathbf{x} \in \mathbb{R}^m : A(\mathbf{x}) \succeq 0\}$ where $A(\mathbf{x}) = A_0 + \sum_{i=1}^m x_i A_i$. If $\hat{\mathbf{x}}$ is an extreme point of S and r is the rank of $A(\hat{\mathbf{x}})$, then*

$$\binom{r+1}{2} + m \leq \binom{N+1}{2}.$$

Furthermore, for generic $A_0, \ldots, A_m$, additionally $m \geq \binom{N-r+1}{2}$.

PROOF. Suppose that $\hat{\mathbf{x}}$ is an extreme point of S and r is the rank of $A(\hat{\mathbf{x}})$. Let L denote the kernel of $A(\hat{\mathbf{x}})$, which has dimension $N - r$. Then the matrix

$A(\hat{\mathbf{x}})$ belongs to the relative interior of the face $\mathcal{F}_L$ of $\mathcal{S}_+^N$. The intersection of $\mathcal{F}_L$ with the affine space parametrized by $A(\mathbf{x})$ is a face of S containing $A(\hat{\mathbf{x}})$, meaning that it must be just one point. Since $\mathcal{F}_L$ is linearly isomorphic to $\mathcal{S}_+^r$, a dimension count then shows that

$$\begin{aligned} m &= \dim(\{A(\mathbf{x}) : \mathbf{x} \in \mathbb{R}^m\}) \ \le\ \operatorname{codim}(\mathcal{F}_L) \\ &= \dim(\mathcal{S}^N) - \dim(\mathcal{F}_L) \ =\ \binom{N+1}{2} - \binom{r+1}{2}. \end{aligned}$$

For the second statement, note that the set $\mathcal{V}_r$ of matrices of rank $\le r$ is a real variety of codimension $\binom{N-r+1}{2}$ in $\mathcal{S}^N$. In particular, a generic affine space of dimension $m < \binom{N-r+1}{2}$ does not intersect $\mathcal{V}_r$. Therefore for generic $A_0, \ldots, A_m$ where $m < \binom{N-r+1}{2}$, there is no $\mathbf{x} \in \mathbb{R}^m$ for which $A(\mathbf{x})$ has rank $\le r$. □

For $N = 4$ and $m = 3$, this states that the rank r of any extreme point satisfies $\binom{r+1}{2} \le \binom{4+1}{2} - 3 = 7$. This holds for $r = 1, 2, 3$, which tells us only that the 4×4 matrix is not full rank. For generic matrices, we find additionally that $3 \ge \binom{5-r}{2}$, which implies that $r \ge 2$. In this case the Pataki range is $\{2, 3\}$.

For higher m, the upper bound on r becomes more restrictive. If $N = 4$ and $m = 5$, then $\binom{r+1}{2} \le \binom{4+1}{2} - 5 = 5$, giving $r = 1, 2$. In this case, there are *never* extreme points of corank one. Indeed, an asymptotic view shows that this is often the case. The dimension of $\mathcal{S}^N$ as a real vector space is $\binom{N+1}{2}$, which is the natural upper bound on m. If we consider $m \approx \frac{1}{2}\binom{N+1}{2}$, then the given upper bound on $\binom{r+1}{2}$ is $\approx \frac{1}{2}\binom{N+1}{2}$, which translates to an upper bound on r of $\approx N/\sqrt{2}$. Therefore the top rank of an extreme point can be much less than $N - 1$.

Example 1.3 cont'd. For our running example, we have $N = 4$ and $m = 3$. There is a curve segment worth of extreme points, $\mathbf{x} = (\cos(\theta), \cos(2\theta), \cos(3\theta))$ for $\theta \in [0, \pi]$. In the interior of this curve, for $\theta \in (0, \pi)$, the corresponding matrix has rank two and at the two endpoints, $\theta \in \{0, \pi\}$, the matrix has rank one.

These ranks satisfy Pataki's inequalities, but we see that their behavior is non-generic. In particular, there are extreme points of rank one and no extreme points of rank three. Taking a wider view though, this example can provide a good mental image of spectrahedra in higher dimensions – most of the points on the boundary are not extreme points and the spectrahedron is the convex hull of a low-dimensional part of its boundary. However, it still exhibits infinite families of faces. ◇

It is not difficult to show that each rank in the Pataki range appears as the rank of extreme points for some full-dimensional set of matrices $A_0, \ldots, A_m \in \mathcal{S}^N$. However examples achieving all of the ranks simultaneously were only discovered recently by Claus Scheiderer [**17**]. These examples come from *Gram spectrahedra* of sums of squares, which provide a fruitful class of examples on which to explore Pataki's inequalities.

4.4. Gram spectrahedra of univariate polynomials. Let $f \in \mathbb{R}[t]_{\le 2d}$ be a positive polynomial of degree at most $2d$, and consider the spectrahedron

$$\operatorname{Gram}_+(f) = \left\{ G \in \mathcal{S}_+^{d+1} : f = \begin{pmatrix} 1 & t & t^2 & \ldots & t^d \end{pmatrix} G \begin{pmatrix} 1 & t & t^2 & \ldots & t^d \end{pmatrix}^T \right\}.$$

As first noted by Choi, Lam, and Reznick [**5**], this spectrahedron provides a concise way of encapsulating the representations of the polynomial f as a sum of squares. This will be explored in greater detail in Section 5.

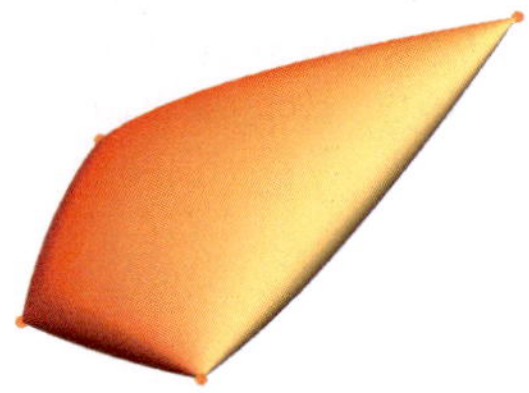

FIGURE 12. The Gram spectrahedron of $1+t^6$ from Example 4.10.

As shown in [**6**], the ranks of the extreme points of $\mathrm{Gram}_+(f)$ are exactly the ranks belonging to the corresponding Pataki range. Each of the $2d+1$ coefficients of f impose linearly independent conditions on the matrix G. If the spectrahedron $\mathrm{Gram}_+(f)$ contains a positive definite matrix, then it has codimension $2d+1$ in $\mathcal{S}_+^{d+1}$ and dimension $m = \binom{d+2}{2} - (2d+1) = \binom{d}{2}$. For a generic positive polynomial $f \in \mathbb{R}[t]_{\leq 2d}$, the ranks r of the extreme points of $\mathrm{Gram}_+(f)$ satisfy

$$\binom{r+1}{2} \leq 2d+1 \quad \text{and} \quad \binom{d}{2} \geq \binom{d+2-r}{2}.$$

This gives that the Pataki interval is $2 \leq r \leq \lfloor(-1+\sqrt{16d+9})/2\rfloor$. The beautiful classical fact that every nonnegative polynomial in $\mathbb{R}[t]$ is a sum of *two* squares shows that the lowest rank two is always achieved. Indeed, if $f = g^2 + h^2$ where $g, h \in \mathbb{R}[t]_{\leq d}$, then we can write $g = \mathbf{m}_d^T \vec{\mathbf{g}}$ and $h = \mathbf{m}_d^T \vec{\mathbf{h}}$, where $\vec{\mathbf{g}}, \vec{\mathbf{h}} \in \mathbb{R}^{d+1}$ are the coefficient vectors of g and h, respectively, and $\mathbf{m}_d^T = \begin{pmatrix} 1 & t & t^2 & \dots & t^d \end{pmatrix}$. The positive semidefinite rank-two matrix $\vec{\mathbf{g}}\vec{\mathbf{g}}^T + \vec{\mathbf{h}}\vec{\mathbf{h}}^T$ belongs to $\mathrm{Gram}_+(f)$ and is an extreme point so long as f is not a single square.

In [**17**], Scheiderer shows that in fact *every* rank in the Pataki range is achieved by an extreme point of such a Gram spectrahedron.

THEOREM 4.9. [**17**] *Let $f \in \mathbb{R}[t]_{\leq 2d}$ be a generic, positive polynomial of degree $2d$. Then the ranks of the extreme points of* $\mathrm{Gram}_+(f)$ *are exactly the ranks belonging to the Pataki range.*

EXAMPLE 4.10 ($d=3$). Consider the univariate polynomial $f = t^6 + 1$. The set of matrices $G \in \mathcal{S}^4$ for which $\mathbf{m}_3^T G \mathbf{m}_3 = f$ is an affine space of dimension $3 = \binom{5}{2} - 7$. Parametrizing this affine linear space by $(a,b,c) \in \mathbb{R}^3$ gives

$$\mathrm{Gram}_+(f) = \left\{ \begin{pmatrix} 1 & 0 & -a & -b \\ 0 & 2a & b & -c \\ -a & b & 2c & 0 \\ -b & -c & 0 & 1 \end{pmatrix} \in \mathcal{S}_+^4 : (a,b,c) \in \mathbb{R}^3 \right\}.$$

It contains four points of rank two given by $(a,b,c) = (0,0,0), \frac{1}{2}(1, \pm\sqrt{3}, 1), (2,0,2)$. The first corresponds to the expected decomposition as a sum of two squares $f = (1)^2 + (t^3)^2$. The last gives the representation $f = (1-2t^2)^2 + (2t-t^3)^2$.

The spectrahedron $\mathrm{Gram}_+(f)$ is shown in Figure 12. As this spectrahedron is not the convex hull of its four extreme points of rank-two, there must also be extreme points of higher rank. Indeed, the remainder of the boundary consists of extreme points of rank three. Therefore the spectrahedron $\mathrm{Gram}_+(f)$ achieves all ranks ($r = 2, 3$) of the Pataki range. ◇

5. Moment problems and sums of squares

One important application of the theory of semidefinite programming is to polynomial optimization, which is the problem of optimizing a polynomial function over a semialgebraic set. In many cases, this can be reformulated as a convex optimization over the convex cone of nonnegative polynomials. For example, to maximize the value of a polynomial $f(\mathbf{x})$ over $\mathbf{x} \in \mathbb{R}^n$, one could instead minimize the value of λ for which the polynomial $\lambda - f(\mathbf{x})$ belongs to the cone of nonnegative polynomials. This cone often proves intractable to work with, and so we consider a cone of "certifiably nonnegative" polynomials, namely sums of squares, in its place. The benefit (and a justification for a chapter on spectrahedra within these lecture notes) is that the cone of sums of squares is the image of the cone of positive semidefinite matrices under a linear map. Therefore one can approximate the solution to a polynomial optimization problem by solving a semidefinite program.

When these approximations are exact (i.e. when they give the correct solution for every objective function), the convex hull of the semialgebraic set being optimized over can be written as a spectrahedral shadow. This is true for many semialgebraic sets and was conjectured to hold for general semialgebraic sets in [**9**]. There are now several counterexamples to this conjecture, known as the Helton-Nie conjecture, discussed further below. These counterexamples come from various convex cones of nonnegative polynomials.

In the real vector space $\mathbb{R}[x_1, \ldots, x_n]_{\leq 2d}$ of real polynomials of degree $\leq 2d$, consider the two convex cones

$$\Sigma_{n,\leq 2d} = \left\{ \sum_{i=1}^{k} h_i^2 \ : \ k \in \mathbb{N}, \quad h_i \in \mathbb{R}[x_1, \ldots, x_n]_{\leq d} \right\}, \text{ and}$$
$$P_{n,\leq 2d} = \{ f \in \mathbb{R}[x_1, \ldots, x_n]_{\leq 2d} \ : \ f(\mathbf{p}) \geq 0 \text{ for all } \mathbf{p} \in \mathbb{R}^n \}.$$

The cone $\Sigma_{n,\leq 2d}$ of sums of squares is used as a tractable inner approximation of $P_{n,\leq 2d}$. Moreover, one can realize this as a projection of $\mathcal{S}_+^N$.

PROPOSITION 5.1. *The convex cone $\Sigma_{n,\leq 2d}$ is the image of $\mathcal{S}_+^N$ under the map*

$$X \mapsto \mathbf{m}_d^T X \mathbf{m}_d,$$

where the entries of $\mathbf{m}_d$ form a basis for $\mathbb{R}[x_1, \ldots, x_n]_{\leq d}$ over $\mathbb{R}$ and $N = \binom{n+d}{d}$.

PROOF. Suppose that $f \in \mathbb{R}[x_1, \ldots, x_n]_{\leq 2d}$ is a sum of squares $f = \sum_{i=1}^k g_i^2$ where $g_i \in \mathbb{R}[x_1, \ldots, x_n]_{\leq d}$. We can write each polynomial as $g_i = \vec{\mathbf{g}}_i^T \mathbf{m}_d$ for some real vector $\vec{\mathbf{g}}_i \in \mathbb{R}^N$. Consider the positive semidefinite matrix $G = \sum_{i=1}^k \vec{\mathbf{g}}_i \vec{\mathbf{g}}_i^T$ in $\mathcal{S}^N$. Then

$$\mathbf{m}_d^T G \mathbf{m}_d \ = \ \sum_{i=1}^{k} \mathbf{m}_d^T \vec{\mathbf{g}}_i \vec{\mathbf{g}}_i^T \mathbf{m}_d \ = \ \sum_{i=1}^{k} (\vec{\mathbf{g}}_i^T \mathbf{m}_d)^2 \ = \ \sum_{i=1}^{k} g_i^2,$$

showing that $f = \sum_{i=1}^k g_i^2$ belongs to the image of $\mathcal{S}_+^N$.

Conversely, suppose $f = \mathbf{m}_d^T G \mathbf{m}_d$ for some positive semidefinite matrix G in $\mathcal{S}_+^N$. We can write $G = \sum_{i=1}^k \vec{\mathbf{g}}_i \vec{\mathbf{g}}_i^T$ for some vectors $\vec{\mathbf{g}}_i \in \mathbb{R}^N$. Then the above equation shows that f is a sum of squares $\sum_i g_i^2$ where $g_i = \vec{\mathbf{g}}_i^T \mathbf{m}_d$. □

By definition $P_{n,\leq 2d}$ is the cone of polynomials on which all linear functions $\{\mathrm{ev}_{\mathbf{p}} : \mathbf{p} \in \mathbb{R}^n\}$ are nonnegative, where for $f \in \mathbb{R}[x_1, \ldots, x_n]_{\leq 2d}$, $\mathrm{ev}_{\mathbf{p}}(f) = f(\mathbf{p})$.

Therefore the cone $P_{n,\leq 2d}$ is dual to the conical hull of the set of these point evaluations. Taking duals again, we find that

$$P^*_{n,\leq 2d} = \overline{\text{conicHull}\left\{\text{ev}_{\mathbf{p}} \in \mathbb{R}[x_1,\ldots,x_n]^*_{\leq 2d} : \mathbf{p} \in \mathbb{R}^n\right\}}.$$

By a classical theorem of Haviland [8], this can also be seen as the set of linear functions $f \mapsto \int f\, d\mu$ obtained by integration with respect to a nonnegative Borel measure on $\mathbb{R}^n$. In order to specify this linear function, it suffices to give its values on the monomial basis $\{x_1^{\alpha_1}\cdots x_n^{\alpha_n} : \alpha \in \mathbb{N}^n, |\alpha| \leq 2d\}$ of $\mathbb{R}[x_1,\ldots,x_n]_{\leq 2d}$. Here $|\alpha|$ denotes $\alpha_1 + \ldots + \alpha_n$. Therefore we can identify $P^*_{n,\leq 2d}$ with the cone of *moments*

$$\begin{aligned} P^*_{n,\leq 2d} &\cong \overline{\left\{\left(\int x_1^{\alpha_1}\cdots x_n^{\alpha_n}\, d\mu\right)_{|\alpha|\leq 2d} : \mu \text{ a nonnegative Borel measure on } \mathbb{R}^n\right\}} \\ &= \overline{\text{conicHull}\left\{(p_1^{\alpha_1}\cdots p_n^{\alpha_n})_{|\alpha|\leq 2d} \ : \ \mathbf{p} \in \mathbb{R}^n\right\}}. \end{aligned}$$

Since duality reverses inclusion, the dual cone to the sums of squares cone is an *outer* approximation of this cone

$$P^*_{n,\leq 2d} \subseteq \Sigma^*_{n,\leq 2d}.$$

Since the sums of squares cone is the image of the cone of positive semidefinite matrices under a linear map, it follows from Proposition 3.3 that $\Sigma^*_{n,\leq 2d}$ is a spectrahedron. If the nonnegative measure μ is a probability measure, that is $\int 1\, d\mu = 1$, then the linear function $\ell \in P^*_{n,\leq 2d}$ given by $\ell(f) = \int f d\mu$ is the *expectation* of f. Elements of the cone $\Sigma^*_{n,\leq 2d}$ that take value one on the constant polynomial $1 \in \mathbb{R}[x_1,\ldots,x_n]_{\leq 2d}$ are sometimes called *pseudo-expectations*.

Every nonnegative polynomial in one variable is a sum of squares, which means that the dual cones $\Sigma^*_{1,\leq 2d}$ and $P^*_{1,\leq 2d}$ are equal.

EXAMPLE 5.2. ($n = 1$, $d = 2$) Choosing the basis $\{1, t, t^2, t^3, t^4\}$ for $\mathbb{R}[t]_{\leq 4}$ identifies the point evaluation ev_p for $p \in \mathbb{R}$ with the point $(1, p, p^2, p^3, p^4) \in \mathbb{R}^5$ and identifies the cone $P^*_{1,\leq 2d}$ with the conic hull of these points. On the other hand, the cone $\Sigma_{1,\leq 4}$ is the image of the cone $\mathcal{S}^3_+$ under the linear map

$$X \mapsto \begin{pmatrix}1 & t & t^2\end{pmatrix} X \begin{pmatrix}1 & t & t^2\end{pmatrix}^T = \sum_{k=0}^{4} \langle A_k, X\rangle t^k$$

where A_k is the real symmetric matrix with entries $(A_k)_{ij} = 1$ for $i+j-2 = k$ and $(A_k)_{ij} = 0$ otherwise. The dual cone $\Sigma^*_{1,\leq 4}$ is then linearly isomorphic to the intersection of the linear space $\text{span}_{\mathbb{R}}\{A_0,\ldots,A_4\}$ and the positive semidefinite cone $\mathcal{S}^3_+$. All together this represents the conic hull of point evaluations as a spectrahedron:

$$\overline{\text{conicHull}\{(1,p,p^2,p^3,p^4) : p \in \mathbb{R}\}} = \left\{(y_0,\ldots,y_4) \in \mathbb{R}^5 : \begin{pmatrix} y_0 & y_1 & y_2 \\ y_1 & y_2 & y_3 \\ y_2 & y_3 & y_4 \end{pmatrix} \succeq 0\right\}.$$

◇

From the equality of $\Sigma^*_{1,\leq 2d}$ and $P^*_{1,\leq 2d}$, we immediately get that the convex hull of any curve parametrized by univariate polynomials is a spectrahedral shadow.

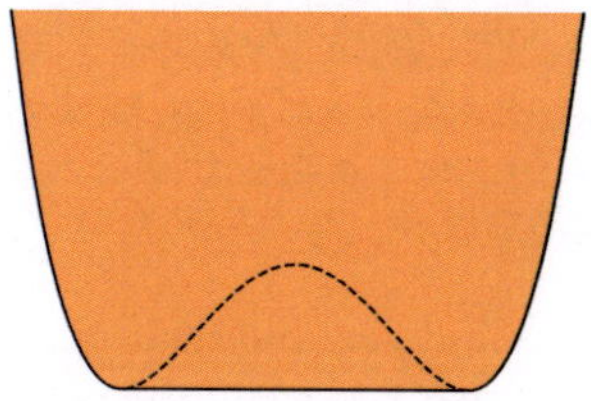

FIGURE 13. The convex hull of the curve in Example 5.3.

EXAMPLE 5.3. Consider the curve in $\mathbb{R}^2$ parametrized by $\gamma(t) = (t, t^4 - 2t^2 + 1)$. The convex hull of this curve is shown in Figure 13. This set is not a spectrahedron, as it has a non-exposed face, but it is a spectrahedral shadow:

$$\operatorname{conv}\{\gamma(t) : t \in \mathbb{R}\} = \left\{ (a, b) \in \mathbb{R}^2 : \exists y_2, y_3 \in \mathbb{R} \text{ s.t. } \begin{pmatrix} 1 & a & y_2 \\ a & y_2 & y_3 \\ y_2 & y_3 & b + 2y_2 - 1 \end{pmatrix} \succeq 0 \right\}.$$

To see this, note that the convex hull $\operatorname{conv}\{\gamma(t) : t \in \mathbb{R}\}$ equals is image of the convex hull $\operatorname{conv}\{(1, t, t^2, t^3, t^4) : t \in \mathbb{R}\} = \Sigma^*_{1,\leq 4} \cap \{y_0 = 1\}$ under the affine linear map $(1, y_1, y_2, y_3, y_4) \mapsto (y_1, y_4 - 2y_2 + 1)$. Solving $y_1 = a$ and $b = y_4 - 2y_2 + 1$ for y_1 and y_4 gives the representation above. ◇

In fact, sums of squares give a good approximation for nonnegative polynomials on any curve. Considering linear polynomials that are nonnegative on a variety as sums of squares gives the following:

THEOREM 5.4 ([**15**]). *The convex hull of any real algebraic curve is a spectrahedral shadow.*

This was originally evidence for the Helton-Nie conjecture that every convex semialgebraic set could be written as the projection of a spectrahedron. This conjecture was disproved by Scheiderer in 2016.

THEOREM 5.5 ([**16**]). *Not every convex, semialgebraic set is a spectrahedral shadow. In particular, the set of nonnegative polynomials of degree ≤ 6 in three variables, $P_{3,\leq 6}$ is not a spectrahedral shadow.*

Building off of these techniques, Hamza Fawzi showed that $P_{2,\leq 6}$ and $P_{3,\leq 4}$ are also not spectrahedral shadows [**7**]. From this, one can show that the cone $P_{n,\leq 2d}$ is not a spectrahedral shadow for all $(n, 2d)$ with $n \geq 2$ and $2d \geq 6$ or with $n \geq 3$ and $2d \geq 4$. Moreover, these are the minimal cases of $(n, 2d)$ in which this is possible.

For $d = 2$, $n = 1$, and $(n, 2d) = (2, 4)$, a classical theorem of Hilbert states that every nonnegative polynomial in $\mathbb{R}[x_1, \ldots, x_n]_{\leq 2d}$ can be written as a sum of squares, i.e. $\Sigma_{n,\leq 2d} = P_{n,\leq 2d}$. By Proposition 5.1, $\Sigma_{n,\leq 2d}$ is the image of $\mathcal{S}^N_+$ under a linear map. Therefore in the Hilbert cases, it follows that the cone $\Sigma_{n,\leq 2d} = P_{n,\leq 2d}$ is the image of $\mathcal{S}^N_+$ under a linear map. Using Proposition 3.3, it follows that $\Sigma^*_{n,\leq 2d} = P^*_{n,\leq 2d}$ is a spectrahedron. All together, this provides a complete characterization of the pairs $(n, 2d)$ for which the cone of nonnegative polynomials $P_{n,\leq 2d}$ is a spectrahedron or spectrahedral shadow.

PROPOSITION 5.6. *For $n, d \in \mathbb{Z}_+$, consider the cones $P_{n,\leq 2d}$ of nonnegative polynomials in $\mathbb{R}[x_1, \ldots, x_n]_{\leq 2d}$ and its dual cone $P^*_{n,\leq 2d}$. Then*

- *both $P_{n,\leq 2}$ and $P^*_{n,\leq 2}$ are spectrahedra,*
- *for $n = 1$ and $(n, 2d) = (2, 4)$, $P^*_{n,\leq 2d}$ is a spectrahedron and $P_{n,\leq 2d}$ is a spectrahedral shadow but not a spectrahedron, and*
- *in all other cases, neither $P_{n,\leq 2d}$ nor $P^*_{n,\leq 2d}$ is a spectrahedral shadow.*

PROOF. For the first, notice that any quadratic polynomials $q \in \mathbb{R}[x_1, \dots, x_n]_{\leq 2}$ can be uniquely represented as $(1, \mathbf{x})Q(1, \mathbf{x})^T$ for a real symmetric matrix $Q \in \mathcal{S}^{n+1}$. Moreover q is nonnegative if and only if Q is positive semidefinite. Therefore $P_{n,\leq 2}$ is linearly isomorphic to the cone $\mathcal{S}^{n+1}_+$, and by the self-duality of $\mathcal{S}^{n+1}_+$, so is $P^*_{n,\leq 2}$.

For $n = 1$ and $(n, 2d) = (2, 4)$, the cones $\Sigma_{n,\leq 2d}$ and $P_{n,\leq 2d}$ are equal. As explained in detail above for $(n, d) = (1, 2)$, the cone $\Sigma_{n,\leq 2d}$ is the image of the positive semidefinite cone under a linear map, meaning that the dual cone $\Sigma^*_{n,\leq 2d} = P^*_{n,\leq 2d}$ is spectrahedral. For $n \geq 1$ and $d \geq 2$, the cone $P_{1,\leq 4}$ is linearly isomorphic to a face of $P_{n,\leq 2d}$, so Example 4.7 shows that $P_{n,\leq 2d}$ has a non-exposed face and therefore cannot be written as a spectrahedron.

Finally the results of Scheiderer [**16**] and Fawzi [**7**] show that in all other cases, $P_{n,\leq 2d}$ is not a spectrahedral shadow. Because the class of spectrahedral shadows is closed under duality, as seen in Proposition 3.4, neither is the dual cone $P^*_{n,\leq 2d}$. □

References

[1] Alexander Barvinok, *A course in convexity*, Graduate Studies in Mathematics, vol. 54, American Mathematical Society, Providence, RI, 2002. MR1940576

[2] Grigoriy Blekherman, Pablo A. Parrilo, and Rekha R. Thomas (eds.), *Semidefinite optimization and convex algebraic geometry*, MOS-SIAM Series on Optimization, vol. 13, Society for Industrial and Applied Mathematics (SIAM), Philadelphia, PA; Mathematical Optimization Society, Philadelphia, PA, 2013. MR3075433

[3] Jacek Bochnak, Michel Coste, and Marie-Françoise Roy, *Real algebraic geometry*, Ergebnisse der Mathematik und ihrer Grenzgebiete (3) [Results in Mathematics and Related Areas (3)], vol. 36, Springer-Verlag, Berlin, 1998. Translated from the 1987 French original; Revised by the authors. MR1659509

[4] Arne Brøndsted, *An introduction to convex polytopes*, Graduate Texts in Mathematics, vol. 90, Springer-Verlag, New York-Berlin, 1983. MR683612

[5] M. D. Choi, T. Y. Lam, and B. Reznick, *Sums of squares of real polynomials*, K-theory and algebraic geometry: connections with quadratic forms and division algebras (Santa Barbara, CA, 1992), Proc. Sympos. Pure Math., vol. 58, Amer. Math. Soc., Providence, RI, 1995, pp. 103–126. MR1327293

[6] Lynn Chua, Daniel Plaumann, Rainer Sinn, and Cynthia Vinzant, *Gram spectrahedra*, Ordered algebraic structures and related topics, Contemp. Math., vol. 697, Amer. Math. Soc., Providence, RI, 2017, pp. 81–105, DOI 10.1090/conm/697/14047. MR3716067

[7] H. Fawzi, *The set of separable states has no finite semidefinite representation except in dimension* 3×2, preprint available at `arXiv:1905.02575`, 2019.

[8] E. K. Haviland, *On the Momentum Problem for Distribution Functions in More Than One Dimension. II*, Amer. J. Math. **58** (1936), no. 1, 164–168, DOI 10.2307/2371063. MR1507139

[9] J. William Helton and Jiawang Nie, *Sufficient and necessary conditions for semidefinite representability of convex hulls and sets*, SIAM J. Optim. **20** (2009), no. 2, 759–791, DOI 10.1137/07070526X. MR2515796

[10] Murray Marshall, *Positive polynomials and sums of squares*, Mathematical Surveys and Monographs, vol. 146, American Mathematical Society, Providence, RI, 2008. MR2383959

[11] Jiawang Nie, Kristian Ranestad, and Bernd Sturmfels, *The algebraic degree of semidefinite programming*, Math. Program. **122** (2010), no. 2, Ser. A, 379–405, DOI 10.1007/s10107-008-0253-6. MR2546336

[12] Gábor Pataki, *On the rank of extreme matrices in semidefinite programs and the multiplicity of optimal eigenvalues*, Math. Oper. Res. **23** (1998), no. 2, 339–358, DOI 10.1287/moor.23.2.339. MR1626662

[13] Motakuri Ramana and A. J. Goldman, *Some geometric results in semidefinite programming*, J. Global Optim. **7** (1995), no. 1, 33–50, DOI 10.1007/BF01100204. MR1342934

[14] Kristian Ranestad, *Algebraic degree in semidefinite and polynomial optimization*, Handbook on semidefinite, conic and polynomial optimization, Internat. Ser. Oper. Res. Management Sci., vol. 166, Springer, New York, 2012, pp. 61–75, DOI 10.1007/978-1-4614-0769-0_3. MR2894691

[15] Claus Scheiderer, *Semidefinite representation for convex hulls of real algebraic curves*, SIAM J. Appl. Algebra Geom. **2** (2018), no. 1, 1–25, DOI 10.1137/17M1115113. MR3755651

[16] Claus Scheiderer, *Spectrahedral shadows*, SIAM J. Appl. Algebra Geom. **2** (2018), no. 1, 26–44, DOI 10.1137/17M1118981. MR3755652

[17] C. Scheiderer, *Extreme points of gram spectrahedra of binary forms*, preprint available at `arXiv:1802.05513`, 2018.

[18] Lieven Vandenberghe and Stephen Boyd, *Semidefinite programming*, SIAM Rev. **38** (1996), no. 1, 49–95, DOI 10.1137/1038003. MR1379041

Department of Mathematics, North Carolina State University, Raleigh, North Carolina

Proceedings of Symposia in Applied Mathematics
Volume **77**, 2020
https://doi.org/10.1090/psapm/077/00682

Lifts of convex sets

Hamza Fawzi

ABSTRACT. A central question in optimization is to maximize (or minimize) a linear function on a given convex set. Such a problem may be easy or hard depending on the geometry of the convex set. This chapter considers the following question: given a convex set, is it possible to express it as the projection of a simpler convex set in a higher-dimensional space? Such a lift potentially allows us to reformulate the original optimization problem as an easier one over the higher-dimensional convex set. We focus in this chapter on two classes of lifts, namely polyhedral and spectrahedral lifts, where a natural notion of complexity can be defined. For spectrahedral lifts, we will see that the existence of lifts is characterized by the existence of sum-of-squares certificates for a certain class of nonnegative functions.

1. Introduction

In this short introductory section we give a brief overview of the lifting method and discuss its importance in optimization when combined with the idea of *convex reformulations*. We illustrate the power of the lifting method on a simple example and introduce linear programming and semidefinite programming lifts.

Convex reformulations. Consider the following problem where we want to minimize a linear function $\ell(x)$ subject to the constraint $x \in X$:

$$\text{minimize} \quad \ell(x) \quad \text{subject to} \quad x \in X. \tag{1}$$

Here X is an arbitrary subset of $\mathbb{R}^n$ and need not be convex. It is a well-known fact that, since the objective function is linear, the optimal value of (1) remains unchanged if we change the constraint "$x \in X$" to "$x \in \text{conv}(X)$", where $\text{conv}(X)$ denotes the convex hull of X (see Figure 1 for an illustration):

$$\text{minimize} \quad \ell(x) \quad \text{subject to} \quad x \in \text{conv}(X). \tag{2}$$

Recall that the convex hull of X is the set of all possible convex combinations of elements of X:

$$\text{conv}(X) = \left\{ \sum_{i=1}^{m} \lambda_i x_i : m \in \mathbb{N}, \lambda_1, \ldots, \lambda_m \geq 0, \sum_{i=1}^{m} \lambda_i = 1, x_1, \ldots, x_m \in X \right\}.$$

This chapter was prepared for the lectures on *Lifts of convex sets* part of the AMS Short Course on Sum of Squares: Theory and Practice, Baltimore, January 14-15, 2019. It contains adapted material from Chapters 1 and 2 of the author's PhD thesis "Power and limitations of convex formulations via linear and semidefinite programming lifts",

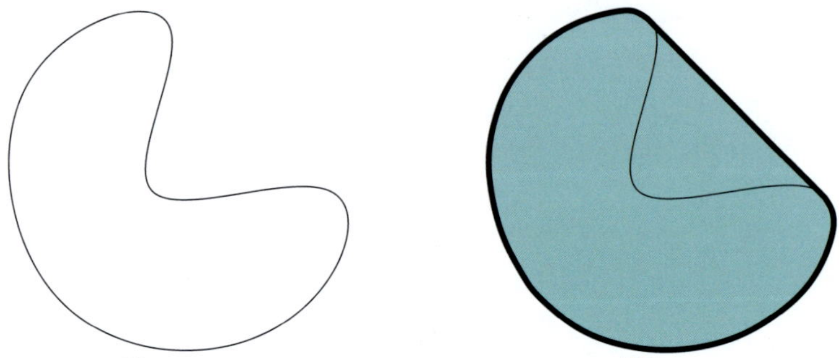

FIGURE 1. A nonconvex set X and its convex hull

Using this definition the equality of the optimal values of (1) and (2) is straightforward to verify. Note that problem (2) is now formally *convex* since the cost function is linear and the feasible set is convex.

The transformation from (1) to (2) seems to rely heavily on the fact that ℓ is a linear function. It turns out however that a similar transformation can be applied more generally, even if the objective is not linear, by introducing additional variables and constraints. To illustrate this, assume that our objective in (1) was quadratic instead of being linear, i.e., we are interested in minimizing $q(x) = \sum_{i\leq j} q_{ij}x_ix_j$ subject to $x \in X$:

$$\text{minimize} \quad q(x) = \sum_{i\leq j} q_{ij}x_ix_j \quad \text{subject to} \quad x \in X. \tag{3}$$

If we introduce additional variables y_{ij} playing the role of x_ix_j we can reformulate the problem above as follows:

$$\text{minimize} \quad \sum_{i\leq j} q_{ij}y_{ij} \quad \text{subject to} \quad y \in Y \tag{4}$$

where Y is defined as

$$Y = \{(y_{ij})_{i\leq j} : \exists x \in X \text{ s.t. } y_{ij} = x_ix_j \ \forall 1 \leq i \leq j \leq n\}.$$

The objective function of (4) is now linear and thus by the same reasoning as above the constraint "$y \in Y$" can be changed to "$y \in \text{conv}(Y)$". In both cases we have transformed the original problem to a new problem that is convex, at least formally. In order to solve the problem however we need to find a tractable representation of the set $\text{conv}(X)$ (or $\text{conv}(Y)$).

REMARK 1. There are different ways one can perform the reformulation above to get a linear objective function and a convex constraint. For example, starting from (3), one could instead introduce a new scalar variable t for the whole quadratic form $\sum_{i\leq j} q_{ij}x_ix_j$. This gives

$$\text{minimize} \quad t \quad \text{subject to} \quad (x,t) \in Z \tag{5}$$

where Z is defined as

$$Z = \left\{ (x,t) \in X \times \mathbb{R} \ : \ \sum_{i\leq j} q_{ij}x_ix_j \leq t \right\}.$$

Since the objective function in (5) is linear one can, in the same way as above, change the constraint $(x, t) \in Z$ to $(x, t) \in \text{conv}(Z)$.

Lifts of convex sets. When the set X is finite (e.g., in discrete and combinatorial optimization), the corresponding convex set $P = \text{conv}(X)$ is called a *polytope* and can be described using a finite number of linear inequalities. The problem of optimizing a linear function over a polytope P is known as linear programming. Interior-point methods are a popular class of algorithms for linear programming, and the complexity of such algorithms typically depend on the size of the inequality description of the polytope.[1] The size of the trivial such description is equal to the number of *facets* of P, a geometric quantity associated to P. Unfortunately, in many cases of interest, the number of facets of P is prohibitively large to enumerate directly in a linear programming formulation.

The idea of *lifting* consists in expressing the polytope P as the projection of a higher-dimensional polytope Q that has much fewer facets than P. We say in this case that Q is a (linear programming) *lift* of P. For the purpose of optimization one can then work over Q rather than working over P. Indeed if we are interested in minimizing ℓ over P and if $P = \pi(Q)$ where π is a linear (projection) map then we have:

$$\min_{x \in P} \ell(x) = \min_{y \in Q} \ell \circ \pi(y). \tag{6}$$

To give a simple example of a lift let P denote the ℓ_1 ball in $\mathbb{R}^n$:

$$P = \{x \in \mathbb{R}^n : \|x\|_1 \leq 1\}.$$

The trivial description of P obtained by enumerating the facets has size 2^n since the facets of the ℓ_1 ball are given by all possible inequalities of the form $\pm x_1 \pm x_2 \pm \cdots \pm x_n \leq 1$, see Figure 2. It is not difficult however to see that the ℓ_1 ball admits a description using only $2n$ linear inequalities, namely:

$$P = \left\{ x \in \mathbb{R}^n : \exists y \in \mathbb{R}^n \text{ s.t. } - y_i \leq x_i \leq y_i, \sum_{i=1}^n y_i = 1 \right\}. \tag{7}$$

In this case the higher-dimensional polytope Q lives in $\mathbb{R}^{2n}$ and consists of vectors (x, y) that satisfy the constraints $-y_i \leq x_i \leq y_i$ and $\sum_{i=1}^n y_i = 1$. The projection map is $\pi : (x, y) \mapsto x$. The key point to note here is that the description (7) is much smaller: we have $2n$ linear inequalities, and 1 linear equality, instead of the 2^n linear inequalities we had before. Observe that the description (7) is the one that we often use in practice when solving ℓ_1 optimization problems. Later we will give other examples of polytopes where lifting allows us to get a description that is much smaller than the trivial one.

We have discussed lifts whereby the polytope P is expressed as the projection of another polytope Q. What if the convex set to start with is not a polytope? There is a more general class of lifts known as *semidefinite programming* lifts. Here the goal is to express a convex set C as the projection of the feasible set of a

[1] We note here that there is another algorithm for linear programming, the *ellipsoid method*, whose complexity does not depend explicitly on the inequality description of the polytope, but on a separation oracle. Thus if P does not have a small inequality description, this does not necessarily mean that we cannot efficiently optimize linear functions over P. A famous example is the matching polytope (see later) which has an efficient separation oracle [**24**], and yet has no polynomial-size linear programming description [**35**].

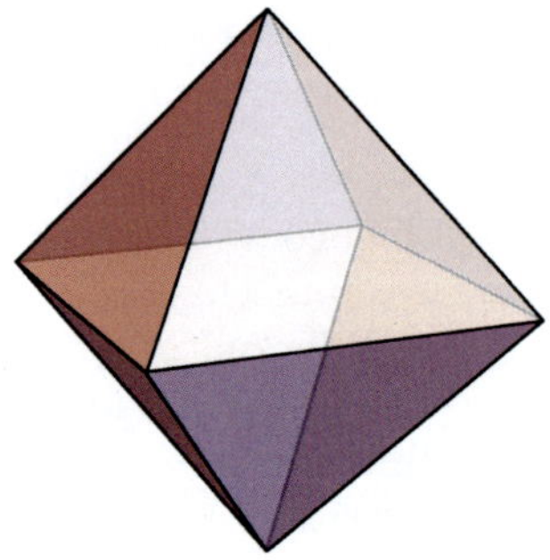

FIGURE 2. The ℓ_1 ball in $\mathbb{R}^3$ has 8 facets – it is an octahedron. The ℓ_1 ball in $\mathbb{R}^n$ has 2^n facets.

semidefinite program (SDP), i.e., a convex set Q that can be described using linear matrix inequalities:

$$Q = \{y \in \mathbb{R}^m : F_0 + y_1 F_1 + \cdots + y_m F_m \succeq 0\}. \tag{8}$$

$F_0, \ldots, F_m$ are $d \times d$ real symmetric matrices and the constraint in (8) indicates that the matrix $F_0 + y_1 F_1 + \cdots + y_m F_m$ is positive semidefinite. Such a convex set is called a *spectrahedron*. When the matrices $F_0, \ldots, F_m$ are diagonal, the set Q is a polytope; in general, however, Q is not necessarily a polytope. It is clear from this observation that SDP lifts form a broader class than LP lifts, even when C is a polytope (see e.g., Figure 6 for a SDP lift of a polytope). One of the main results we will prove is a connection between semidefinite lifts and sums of squares. More precisely we will prove in Theorem 5 that a convex set has a small semidefinite lift if and only if all its valid linear inequalities admit sum-of-squares certificates from a *low-dimensional* subspace.

History. The idea of lifting (also known as *extended formulations*) is old in optimization, and can be traced back to the early work of Balas on disjunctive programming [**2**]. See [**8**] for a modern account, and for connections and applications in combinatorial optimization. The first paper that studies lifts in a systematic way, proving nonexistence of small lifts for certain polytopes, is by Yannakakis in 1991 [**40**]. In his paper Yannakakis showed that the traveling salesman polytope and the matching polytope do not have polynomial-size symmetric linear programming lifts.

The recent years have witnessed a resurgence of interest in this topic. From the lower bounds point of view several results have been proved concerning the nonexistence of small lifts for polytopes arising in combinatorial optimization. Fiorini et al. [**17**] resolved a conjecture left open by Yannakakis and proved that the traveling salesman polytope does not admit any polynomial-size linear programming lifts (without any symmetry requirement). Later, Lee, Raghavendra, Steurer [**30**] showed that the traveling salesman polytope has no polynomial-size semidefinite programming lift. Another major result was also obtained recently by Rothvoß [**35**] where he showed that the matching polytope has no polynomial-size linear programming lift (again, with no symmetry requirement). This result of Rothvoß is particularly striking since the matching polytope is known to have a polynomial-time separation oracle. Several results have also been obtained concerning approximate lifts, see for example [**5–7**].

From the upper bounds point of view new methods have been proposed to construct improved lifts for certain classes of polytopes and convex sets, see for example [**14**, **19**, **27**, **28**, **37**]. The problem of constructing semidefinite programming lifts for algebraic sets has been of specific importance in the area of *convex algebraic geometry* [**4**, **23**, **29**]. In [**25**] Helton and Nie conjectured that any convex semialgebraic set admits a semidefinite programming lift; however, this was recently disproved in [**38**].

2. Linear programming lifts and Yannakakis' theorem

The main goal of this section is to give a concrete way to think about lifts of a polytope in terms of nonnegative factorizations of a certain matrix associated to P. This matrix, known as the *slack matrix*, encodes information about the extreme points and the facet inequalities of P. We first need to recall some basic and important facts about polytopes and linear programming.

Polytopes. A polytope is the convex hull of a finite number of points. Any polytope can also be described using a finite number of linear inequalities

$$P = \{x \in \mathbb{R}^n : Fx \leq g\}. \tag{9}$$

Here $F \in \mathbb{R}^{k\times n}$, $g \in \mathbb{R}^k$ and the inequality $Fx \leq g$ is understood componentwise. The minimal number k of linear inequalities needed is the number of *facets* of P (assuming P does not lie in a low-dimensional affine subspace), see Figure 3. Linear

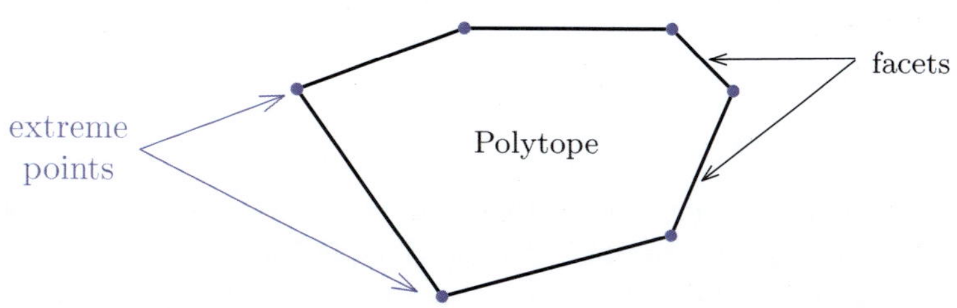

FIGURE 3. A polytope is the convex hull of a finite number of points, called *extreme points* or *vertices*. Any polytope has also a dual description in terms of a finite number of linear inequalities.

programming is the problem of minimizing (or maximizing) a linear function subject to linear inequalities. The standard form of linear programming is

$$\min_{z\in\mathbb{R}^N} \quad c^T z \quad \text{subject to} \quad Az = b,\ z \geq 0. \tag{10}$$

Again, the inequality $z \geq 0$ is understood componentwise. It is not difficult to verify (exercise!) that minimizing a linear function over a polytope can be modeled as a linear program (10) where N (the dimension of the variable z in (10)) is equal to the number of facets of a polytope. A natural question is to ask whether one can do better. This motivates the following definition:

DEFINITION 1. Let P be a polytope in $\mathbb{R}^n$. We say that P has a *LP lift of size d* if P can be written as $P = \pi(\mathbb{R}^d_+ \cap L)$ where $\pi : \mathbb{R}^d \to \mathbb{R}^n$ is a linear map,

$\mathbb{R}^d_+ = \{x \in \mathbb{R}^d : x \geq 0\}$ and L is an affine subspace of $\mathbb{R}^d$. The size of the smallest LP lift of P is called the *LP extension complexity* of P.

If a polytope $P \subset \mathbb{R}^n$ has an LP lift of size d then any linear optimization problem over P can be formulated as a linear program in standard form (10) of size d. Indeed, assume $P = \pi(\mathbb{R}^d_+ \cap L)$. For any $c \in \mathbb{R}^n$ we have

$$\min\left\{c^T x : x \in P\right\} = \min_{z \in \mathbb{R}^d}\left\{\tilde{c}^T z : z \in L, z \geq 0\right\} \tag{11}$$

where $\tilde{c} = \pi^*(c)$. The problem on the right-hand side of (11) is precisely of the form (10) when A, b are chosen to describe the affine subspace L.

One can equivalently define an LP lift as a representation $P = \pi(Q)$ where Q is a polytope with d facets. We leave it as an exercise to the reader to verify this. Figure 4 shows an LP lift of size 5 for the regular hexagon in the plane.

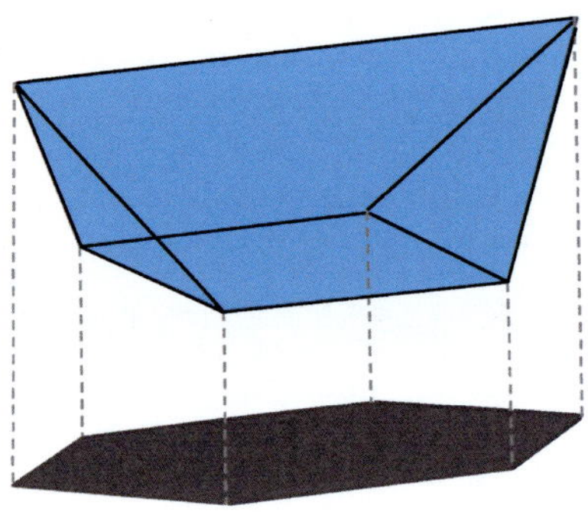

FIGURE 4. LP lift of a hexagon of size 5. Note that hexagon has 6 facets whereas the higher-dimensional polytope has 5 facets in $\mathbb{R}^3$.

REMARK 2. The reader will notice that the size of an LP lift (Definition 1) is only determined by the number of linear inequalities d, and not the number of linear equalities equal to $\operatorname{codim}(L)$. One reason for this is that $\operatorname{codim}(L)$ is always bounded above by d and so we only lose a factor of at most two when using d. Another reason for using d is merely convenience: it gives "cleaner" statements with regards to the connection with nonnegative matrix factorizations, as will be clear later.

Examples of LP lifts. We now give some examples of polytopes P that admit nontrivial lifts.

- We saw earlier the example of the ℓ_1 ball in $\mathbb{R}^n$ which has 2^n facets and which admits a simple lift of size $2n$, see Equation (7).
- Another example of nontrivial lift is for the *permutahedron*. The permutahedron $P \subset \mathbb{R}^n$ is defined as the convex hull of all possible permutations of the vector $(1, 2, \ldots, n)$, i.e.:

  $$P = \operatorname{conv}\left\{(\sigma(1), \ldots, \sigma(n)) : \sigma \in \mathfrak{S}_n\right\}$$

  where $\mathfrak{S}_n$ is the set of permutations on $\{1, \ldots, n\}$. This polytope arises naturally in ordering problems such as in gene sequencing, see e.g. [**31**]. It is known that the permutahedron has an exponential number of facets, precisely $2^n - 2$. However it is not very difficult to construct a lift of the permutahedron of size $\approx n^2$. Indeed let Q denote the convex hull of *permutation matrices* in $\mathbb{R}^{n\times n}$. The Birkhoff-von Neumann theorem asserts that Q is precisely the set of doubly stochastic matrices, i.e.:

$$Q = \Big\{ M \in \mathbb{R}^{n\times n} \;:\; M_{ij} \geq 0 \;\; \forall i,j = 1,\dots,n, \quad \sum_{i=1}^{n} M_{ij} = 1,\; \forall j = 1,\dots,n, \quad \sum_{j=1}^{n} M_{ij} = 1,\; \forall i = 1,\dots,n \Big\}.$$

It is easy to see that P is a projection of Q: indeed if we let $\pi : \mathbb{R}^{n\times n} \to \mathbb{R}^n$ defined by $\pi(M) = Mu$ where $u = (1,2,\dots,n)^T$ then we get that $\pi(Q) = P$. This lift has size n^2 because Q requires exactly n^2 inequalities for its description. It turns out however that this lift is not optimal. Goemans showed in [**19**] that the permutahedron admits a lift of size $O(n \log n)$ which is optimal (i.e., there is no smaller possible lift). His construction however is more complicated and makes use of sorting networks.

- Let $P \subset \mathbb{R}^2$ be the regular N-gon in the plane, i.e., the convex hull of the N complex roots of unity. Even though P has N facets (and N extreme points), a result of Ben-Tal and Nemirovski [**3**] shows that the regular N-gon admits a lifted description with only $O(\log N)$ inequalities. This construction was used in [**3**] to obtain polyhedral approximations of the second-order cone. Figure 4 shows a lift of the regular hexagon of size 5.
- For other examples of lifts from the combinatorial optimization literature we refer the reader to the surveys [**8**, **26**].

Slack matrix. Let P be a polytope and let X be its extreme points so that $P = \operatorname{conv}(X)$. We saw that any polytope can be described by a finite set of linear inequalities, its *facets* (see Figure 3). Any facet has the form $\ell(x) \leq \ell_{\max}$ where ℓ is a linear form and $\ell_{\max} = \max_{x\in X} \ell(x)$. We now introduce the crucial concept of *slack matrix* needed for the statement of Yannakakis' theorem, which combines both the vertex representation and the facet representation of a polytope.

DEFINITION 2 (Slack matrix). Let $P = \operatorname{conv}(X)$ be a polytope in $\mathbb{R}^n$, with X being the set of its extreme points. The *slack matrix* of P is a matrix where rows are indexed by extreme points of P and columns are indexed by facets of P, and is defined as follows: The value at the (x,ℓ) entry of P (where x is an extreme point and $\ell \leq \ell_{\max}$ is a facet inequality of P) is given by:

$$S_{x,\ell} = \ell_{\max} - \ell(x).$$

The slack matrix of P is not, strictly speaking, uniquely defined since it depends on the ordering of the facets/extreme points and the scaling of the facet inequalities. The results stated here however will be independent of the choice of slack matrix and that is why we will often talk about "the" slack matrix of P.

EXAMPLE 1 (Slack matrix of the regular hexagon). To illustrate the definition of the slack matrix consider the case of the regular hexagon in $\mathbb{R}^2$. This polytope has $|X| = 6$ extreme points, as well as 6 facets so its slack matrix is 6×6. Figure 5 (left) gives the slack matrix.

$$(12)\qquad \begin{bmatrix} 0 & 1 & 2 & 2 & 1 & 0 \\ 0 & 0 & 1 & 2 & 2 & 1 \\ 1 & 0 & 0 & 1 & 2 & 2 \\ 2 & 1 & 0 & 0 & 1 & 2 \\ 2 & 2 & 1 & 0 & 0 & 1 \\ 1 & 2 & 2 & 1 & 0 & 0 \end{bmatrix}$$

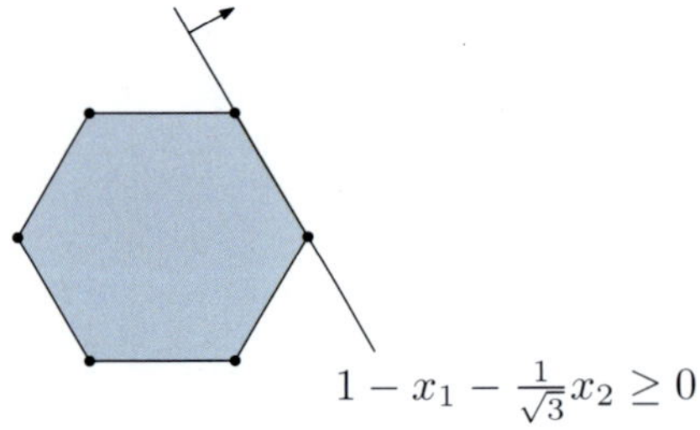

FIGURE 5. Slack matrix of the regular hexagon

For example to see how the first column is formed consider the facet inequality of the hexagon given by $1 - x_1 - x_2/\sqrt{3} \geq 0$. The k'th vertex of the hexagon has coordinates $(\cos(2(k-1)\pi/6), \sin(2(k-1)\pi/6))$. The k'th component of the first column is thus given by $1 - \cos(2(k-1)\pi/6) - \sin(2(k-1)\pi/6)/\sqrt{3}$. Note that each column has exactly two zeros: this is because each facet of the regular hexagon passes through exactly two extreme points.

The entries of a slack matrix are all nonnegative, by definition. We now introduce a second important definition needed for the statement of Yannakakis' theorem.

DEFINITION 3 (Nonnegative matrix factorization / nonnegative rank). Let $S \in \mathbb{R}_+^{p\times q}$ be a matrix with nonnegative entries. We say that S has a *nonnegative factorization* of size r if we can write $S = AB$ where $A \in \mathbb{R}_+^{p\times r}$ and $B \in \mathbb{R}_+^{r\times q}$ have nonnegative entries. The smallest r for which such a factorization exists is called the *nonnegative rank* of S and denoted $\operatorname{rank}_+(S)$.

Note that the *rank* of a matrix $S \in \mathbb{R}^{p\times q}$ is the smallest r such that $S = AB$ where A and B are arbitrary matrices of size $p \times r$ and $r \times q$ respectively. As such, for a nonnegative matrix S we always have $\operatorname{rank}(S) \leq \operatorname{rank}_+(S)$; furthermore, the inequality can be strict (see exercise below).

EXERCISE 1. Write down the slack matrix of the square $[0,1]^2$. What is the rank of this matrix? What is the nonnegative rank?

Yannakakis' theorem. We are now ready to state Yannakakis' theorem which establishes a connection between LP lifts of a polytope, and nonnegative factorizations of its slack matrix.

THEOREM 1 (Yannakakis [**40**]). *Let P be a polytope and let S be its slack matrix. Then P has a LP lift of size d if, and only if, S has a nonnegative factorization of size d. As a consequence, the smallest size of a LP lift of P is equal to* $\operatorname{rank}_+(S)$.

Before presenting the proof of the theorem we discuss an example and an interpretation of Yannakakis' theorem that will be useful later.

EXAMPLE 2 (Regular hexagon continued). One can show that the slack matrix of the regular hexagon has the following nonnegative factorization of size 5:

$$\underbrace{\left[\begin{array}{c|c|c|c|c|c} 0 & 1 & 2 & 2 & 1 & 0 \\ 0 & 0 & 1 & 2 & 2 & 1 \\ 1 & 0 & 0 & 1 & 2 & 2 \\ 2 & 1 & 0 & 0 & 1 & 2 \\ 2 & 2 & 1 & 0 & 0 & 1 \\ 1 & 2 & 2 & 1 & 0 & 0 \end{array}\right]}_{\text{slack matrix of hexagon}} = \left[\begin{array}{c|c|c|c|c} 0 & 1 & 0 & 0 & 1 \\ 0 & 2 & 0 & 1 & 0 \\ 0 & 1 & 1 & 0 & 0 \\ 1 & 0 & 1 & 0 & 0 \\ 2 & 0 & 0 & 1 & 0 \\ 1 & 0 & 0 & 0 & 1 \end{array}\right] \left[\begin{array}{c|c|c|c|c|c} 1 & 1 & 0 & 0 & 0 & 0 \\ 0 & 0 & 0 & 1 & 1 & 0 \\ 1 & 0 & 0 & 0 & 1 & 2 \\ 0 & 0 & 1 & 0 & 0 & 1 \\ 0 & 1 & 2 & 1 & 0 & 0 \end{array}\right]. \tag{13}$$

Note that the matrices on the right-hand side are elementwise nonnegative. By Yannakakis' theorem this factorization says that the regular hexagon has a LP lift of size 5, see Figure 4. One can actually show that the regular hexagon has no smaller LP lift.

Interpretation of Yannakakis' theorem in terms of certificates of nonnegativity. Let P be a polytope and let X be the set of extreme points of the polytope P. If $\ell \leq \ell_{\max}$ is a facet inequality for P then this implies that $\ell_{\max} - \ell|_X$ (i.e., the restriction of $\ell_{\max} - \ell$ to X) is a nonnegative function on X. Yannakakis' theorem can be restated as follows:

THEOREM 2 (Yannakakis [**40**]; restatement of Theorem 1). *Let $P = \operatorname{conv}(X) \subset \mathbb{R}^n$ be a polytope. Then P has an LP lift of size d if and only if, there exist d nonnegative functions on X, $a_1, \dots, a_d : X \to \mathbb{R}_+$ such that the following holds: for any facet inequality $\ell(x) \leq \ell_{\max}$ of P there exist nonnegative coefficients $b_1, \dots, b_d \geq 0$ such that*

$$\ell_{\max} - \ell|_X = \sum_{i=1}^{d} b_i a_i. \tag{14}$$

To make the connection with Theorem 1, the functions $a_1, \dots, a_d : X \to \mathbb{R}_+$ are nothing but the columns of A in a nonnegative factorization of the slack matrix $S = AB$, and the coefficients $b_1, \dots, b_d$ are the entries in the ℓ'th column of B.

Equation (14) is an equality of functions on X. It should be interpreted as a certificate of nonnegativity of $\ell_{\max} - \ell$ on X: indeed the right hand side is a linear combination with nonnegative weights (the b_i's) of nonnegative functions $a_1, \dots, a_d$, and so is "obviously" nonnegative on X. The size of the certificate, here the number of functions $a_1, \dots, a_d$, gives us the size of the LP lift. The main question in constructing a lift is of course to come up with the nonnegative functions $a_1, \dots, a_d$. This point of view will be useful later when we consider semidefinite programming lifts since it will allow us to establish the connection with sums of squares.

Proof of Yannakakis' theorem. We now present a proof of Yannakakis' theorem.

PROOF OF THEOREM 2. We first prove necessity. Assume $P = \pi(\mathbb{R}^d_+ \cap L)$ is an LP lift of P of size d, where L is an affine subspace of $\mathbb{R}^d$ and π a linear map. Let $\ell_{\max} \leq \ell$ be any facet inequality of P. Since $\pi(\mathbb{R}^d_+ \cap L) \subseteq P$ it is clear that the following implication holds for any y:

$$\begin{cases} y \geq 0 \\ y \in L \end{cases} \Rightarrow \ell \circ \pi(y) \leq \ell_{\max}.$$

One can show using Farkas' lemma/strong duality for LP, that there exist coefficients $b_1, \ldots, b_d \geq 0$ and an affine form γ that vanishes on L such that

$$\ell_{\max} - \ell \circ \pi(y) = \sum_{i=1}^{d} b_i y_i + \gamma(y). \tag{15}$$

Note that Equation (15) automatically implies that $\ell_{\max} - \ell \circ \pi$ is nonnegative on $\mathbb{R}^d_+ \cap L$. To see why (15) is true let L_0 be the linear subspace of $\mathbb{R}^d$ parallel to L and let $y_0 \in \mathbb{R}^d$ such that $L = y_0 + L_0$. Consider the following primal/dual pair of LPs whose value is $\ell_{\max}$:[2]

$$\begin{array}{llcll} \max\limits_{y \in \mathbb{R}^d} & (\ell \circ \pi)(y) & \qquad & \min\limits_{b,h \in \mathbb{R}^d} & -\langle h, y_0 \rangle \\ \text{s.t.} & y \in \mathbb{R}^d_+ & & \text{s.t.} & -\ell \circ \pi = b + h \\ & y - y_0 \in L_0 & & & b \in \mathbb{R}^d_+, \; h \in L_0^\perp \end{array} \tag{16}$$

By strong duality, there exists $b \in \mathbb{R}^d_+$ and $h \in L_0^\perp$ such that $-\ell \circ \pi = b + h$ and $-\langle h, y_0 \rangle = \ell_{\max}$. Thus this means that

$$\ell_{\max} - \ell \circ \pi = b + h - \langle h, y_0 \rangle$$

which is exactly (15) with $\gamma(y) = \langle h, y - y_0 \rangle$. Now for any $x \in X$ we know that there exists $A(x) = (a_1(x), \ldots, a_d(x)) \in \mathbb{R}^d_+ \cap L$ such that $\pi(A(x)) = x$. By evaluating Equation (15) at $y = A(x)$ we get (using the fact that $A(x) \in L$ and so $\gamma(A(x)) = 0$):

$$\ell_{\max} - \ell(x) = \sum_{i=1}^{d} b_i a_i(x) \quad \forall x \in X.$$

Thus this proves the claim.

We now show sufficiency. Assume $P = \{x \in \mathbb{R}^n : Fx \leq g\}$ is a facet description of our polytope P where $F \in \mathbb{R}^{N \times n}$ and $g \in \mathbb{R}^N$. Assume $S = AB$ is a nonnegative factorization of the slack matrix of size d, where $A \in \mathbb{R}^{|X| \times d}_+$ and $B \in \mathbb{R}^{d \times N}_+$. Then it is easy to verify that P can be written as:

$$P = \{x \in \mathbb{R}^n : \exists a \in \mathbb{R}^d \text{ s.t. } a \geq 0, \; g - Fx = B^T a\}. \tag{17}$$

To see why the inclusion "$\subseteq$" holds, note that if $x \in X$ is a vertex of P then by letting a be the row of A indexed by x the constraints on the right-hand side are satisfied. The inclusion "$\supseteq$" is trivial since $B^T a \geq 0$ for $a \geq 0$. The proof is almost complete since the right-hand side of (17) is defined using only d linear inequalities. To be sure we just need to show that (17) can be put in the form $P = \pi(\mathbb{R}^d_+ \cap L)$ for some linear map π and affine subspace $L \subset \mathbb{R}^d$. Since P is bounded and nonempty, we know that $\operatorname{rank}(F) = n$ and $g \notin \operatorname{Im}(F)$. Since $\operatorname{rank}(F) = n$ the equation $Fx = g - B^T a$ (in x) has a unique solution x_a if $g - B^T a \in \operatorname{Im}(F)$ and no solution otherwise. It is easy to see that the map that sends a to x_a, defined on the affine subspace $L = \{a \in \mathbb{R}^d : g - B^T a \in \operatorname{Im}(F)\}$, is affine. Since 0 does not belong to L (this is because $g \notin \operatorname{Im}(F)$) this affine map can be extended to a linear map π

[2] In the dual problem's constraint $-\ell \circ \pi = b + h$, we are identifying the vectors b and h with the associated linear forms $x \mapsto b^T x$ and $x \mapsto h^T x$

on the whole space. We thus finally get that $P = \pi(\mathbb{R}^d_+ \cap L)$. This shows that P admits a LP lift of size d. ☐

3. Semidefinite programming lifts, positive semidefinite factorizations, and sums of squares

In this section we extend the concepts and results to *semidefinite programming* lifts.

Semidefinite programming. A *semidefinite program* (SDP) in standard form is an optimization problem of the form:

$$\min_{Y \in \mathbf{S}^d} \langle C, Y \rangle \quad \text{subject to} \quad \mathcal{A}(Y) = b, \;\; Y \in \mathbf{S}^d_+ \tag{18}$$

where $\mathbf{S}^d$ is the space of $d \times d$ real symmetric matrices, $\mathbf{S}^d_+$ is the cone of positive semidefinite matrices in $\mathbf{S}^d$, and $\langle C, Y \rangle = \mathrm{Tr}(CY)$ is the trace inner product of C and Y. A semidefinite program is completely specified by the triple $(C, \mathcal{A}, b)$ where the matrix $C \in \mathbf{S}^d$ specifies the objective function, $\mathcal{A} : \mathbf{S}^d \to \mathbb{R}^m$ is a linear map and $b \in \mathbb{R}^m$. The feasible set of a semidefinite program is called a *spectrahedron* and is an intersection of the positive semidefinite cone $\mathbf{S}^d_+$ with an affine subspace $\{Y : \mathcal{A}(Y) = b\}$. We refer the reader to the chapter [**39**] for more information on the geometry of spectrahedra.

A natural question is to understand which optimization problems can be put in semidefinite programming form. This motivates the following definition of SDP lift:

DEFINITION 4. Let C be a convex set in $\mathbb{R}^n$. We say that C has a SDP lift of size d if it can be written as $C = \pi(\mathbf{S}^d_+ \cap L)$ where $\pi : \mathbf{S}^d \to \mathbb{R}^n$ is a linear map, $\mathbf{S}^d_+$ is the cone of $d \times d$ real symmetric positive semidefinite matrices, and L is an affine subspace of $\mathbf{S}^d$. The size of the smallest SDP lift of C is called the *SDP extension complexity* of C.

If C has an SDP lift of size d, then any linear optimization problem over C can be modeled as an SDP in standard form of size d. Indeed if $C = \pi(\mathbf{S}^d_+ \cap L)$ for some affine subspace L of $\mathbf{S}^d$ and linear map π, then we have:

$$\min\left\{c^T x : x \in C\right\} \;=\; \min\left\{\langle M, Y \rangle : Y \in L, Y \in \mathbf{S}^d_+\right\}$$

where $M = \pi^*(c)$. The right-hand side is an SDP in standard form (18) with an appropriate choice of $\mathcal{A}, b$.

EXAMPLE 3 (Disk). Let $D = \{(x_1, x_2) \in \mathbb{R}^2 : x_1^2 + x_2^2 \leq 1\}$. One can check that D has an SDP lift of size 2, namely $D = \pi(\mathbf{S}^2_+ \cap L)$ where $L = \left\{Y \in \mathbf{S}^2 : \mathrm{Tr}(Y) = 2\right\}$ and $\pi\left(\left[\begin{smallmatrix} a & b \\ b & c \end{smallmatrix}\right]\right) = ((c-a)/2, b)$. We leave it as an exercise to the reader to verify this.

EXAMPLE 4. Figure 6 illustrates a SDP lift of the square $[-1,1]^2$ of size 3, given by:

$$[-1,1]^2 = \left\{ (x_1, x_2) \in \mathbb{R}^2 : \exists u \in \mathbb{R} \begin{bmatrix} 1 & x_1 & x_2 \\ x_1 & 1 & u \\ x_2 & u & 1 \end{bmatrix} \succeq 0 \right\}. \tag{19}$$

Using the notations of Definition 4, $L = \{Y \in \mathbf{S}^3 : Y_{11} = Y_{22} = Y_{33} = 1\}$ and $\pi(Y) = (Y_{12}, Y_{13})$. To see why (19) is true note that if $(x_1, x_2) \in \{-1, 1\}^2$ then by letting $u = x_1 x_2$ we have:

$$\begin{bmatrix} 1 & x_1 & x_2 \\ x_1 & 1 & u \\ x_2 & u & 1 \end{bmatrix} = \begin{bmatrix} 1 \\ x_1 \\ x_2 \end{bmatrix} \begin{bmatrix} 1 \\ x_1 \\ x_2 \end{bmatrix}^T \succeq 0.$$

Since the right-hand side of (19) is convex this shows that $\text{conv}(\{-1, 1\}^2) = [-1, 1]^2$ is contained in it. Conversely if (x_1, x_2) belongs to the right-hand side of (19) then by looking at the 2×2 minors of the 3×3 positive semidefinite matrix we easily get that $1 - x_1^2 \geq 0$ and $1 - x_2^2 \geq 0$ i.e., $(x_1, x_2) \in [-1, 1]^2$.

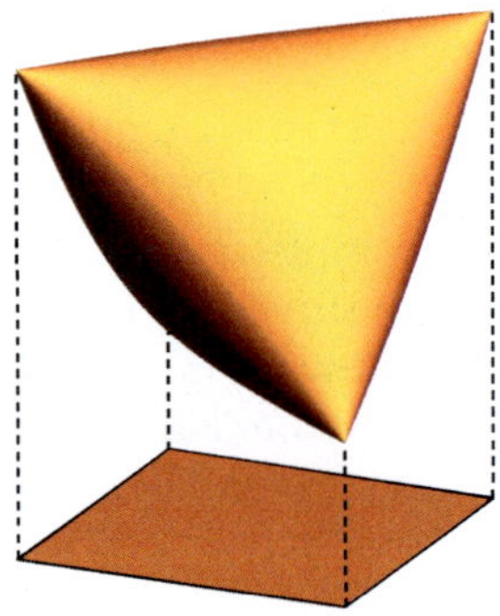

FIGURE 6. SDP lift of the square $[-1, 1]^2$ of size 3 (cf. Equation (19)). The three-dimensional convex set shown in the figure is the set of (x_1, x_2, u) such that the 3×3 symmetric matrix on the right-hand side of (19) is positive semidefinite. Projecting this set onto (x_1, x_2) yields the square $[-1, 1]^2$.

EXERCISE 2. Give an SDP lift of the convex set $C = \{(x_1, x_2) \in \mathbb{R}^2 : x_1^4 + x_2^4 \leq 1\}$.

EXERCISE 3. Give an SDP lift of the convex set $\{A \in \mathbb{R}^{m \times n} : \|A\|_* \leq 1\}$ where $\|A\|_*$ is the *nuclear norm* of A defined as the sum of the singular values of A.

Our goal now is to state a result similar to Yannakakis' theorem to characterize *SDP lifts*. Before we do so, we need to recall some basic facts about general convex sets.

Polars of convex sets and biduality. Let $C \subset \mathbb{R}^n$ be a closed convex set and assume (without loss of generality) that C has non-empty interior and that the origin is in its interior. Any such convex set will be called a *convex body*. The following are two important definitions related to convex sets:

DEFINITION 5 (Extreme point). Let C be a convex set. A point $x \in C$ is called an *extreme point* if whenever $x = (a + b)/2$ with $a, b \in C$ then $a = b = x$. The set of extreme points of C will be denoted extreme(C).

The *Minkowski theorem* (also called the Krein-Milman theorem) says that if C is a closed and bounded convex set in $\mathbb{R}^n$ then C is the convex hull of its extreme points.

DEFINITION 6 (Polar). The *polar* of a convex body C is defined as:

$$C^\circ = \{y \in \mathbb{R}^n : \langle y, x\rangle \leq 1 \; \forall x \in C\}.$$

In words, C° is the set of linear forms $\langle y, \cdot\rangle$ whose maximum on C is at most 1. The *biduality theorem* tells us that these linear forms give a complete inequality description of C, namely:

$$C = \{x \in \mathbb{R}^n : \langle y, x\rangle \leq 1 \; \forall y \in C^\circ\} = (C^\circ)^\circ. \tag{20}$$

One should compare the equation above with the inequality description of a polytope $P = \{x : Fx \leq g\}$. By convexity it is enough to consider $y \in \text{extreme}(C^\circ)$ in Equation (20), i.e., we can write

$$C = \{x \in \mathbb{R}^n : \langle y, x\rangle \leq 1 \; \forall y \in \text{extreme}(C^\circ)\}.$$

If C is a polytope then C° has finitely many extreme points which are in one-to-one correspondence with the facets of C.

Slack matrix of convex sets. In the previous section we defined the slack matrix of a polytope. We are now going to generalize this definition to convex bodies.

DEFINITION 7 (Slack matrix of a convex body). Let $C \subset \mathbb{R}^n$ be a convex body. The slack matrix of C is a (potentially infinite) matrix S whose rows are indexed by the extreme points of C, columns are indexed by the extreme points of C° and is defined as follows:

$$S_{x,y} = 1 - \langle y, x\rangle \quad \forall x \in \text{extreme}(C), y \in \text{extreme}(C^\circ).$$

Note that if C is a polytope then we recover the same definition from the previous section. This is because when C is a polytope the facet inequalities are precisely of the form $\langle y, x\rangle \leq 1$ where $y \in \text{extreme}(C^\circ)$.

Positive semidefinite factorization. We now introduce the notion of positive semidefinite factorization of a nonnegative matrix introduced in [**21**].

DEFINITION 8 (Positive semidefinite factorization [**13, 21**]). Let $S \in \mathbb{R}_+^{I\times J}$ be a matrix with nonnegative entries where rows are indexed by I, columns are indexed by J, and I and J are potentially infinite. We say that S has a *positive semidefinite factorization (psd factorization)* of size d if there exist positive semidefinite matrices $A_i \in \mathbf{S}_+^d$ for each $i \in I$, and $B_j \in \mathbf{S}_+^d$ for each $j \in J$, such that $S_{ij} = \langle A_i, B_j\rangle$ for all $i \in I$ and $j \in J$. The size of the smallest psd factorization of S is called the *psd rank* of S and denoted $\text{rank}_{\text{psd}}(S)$.

EXERCISE 4. Show that for any matrix S we have $\frac{1}{2}\sqrt{1 + 8\,\text{rank}(S)} - \frac{1}{2} \leq \text{rank}_{\text{psd}}(S) \leq \text{rank}_+(S)$.

EXERCISE 5. Let $S \in \mathbb{R}_+^{n\times n}$ defined by $S_{ij} = (i-j)^2$ for $1 \leq i, j \leq n$. Show that $\text{rank}_{\text{psd}}(S) = 2$. Show that $\text{rank}_+(S) \geq \log_2(n)$.

We are now ready to state a result similar to that of Yannakakis to characterize the existence of SDP lifts. It is due to Gouveia, Parrilo, Thomas [**21**].

THEOREM 3 (Gouveia, Parrilo, Thomas [**21**]). *Let C be a convex body and let S be its slack matrix. Then C has a SDP lift of size d if, and only if, S has a positive semidefinite factorization of size d. As a consequence, the smallest size of a SDP lift of P is equal to* $\text{rank}_{\text{psd}}(S)$.

EXAMPLE 5 (Disk). Consider the example of the disk $D = \{(x_1, x_2) : x_1^2 + x_2^2 \leq 1\}$. The polar of D is D itself, $D^\circ = D$. One can verify that the slack matrix of D is the infinite matrix $S_{x,y} = 1 - \langle x, y\rangle$ where rows and columns are indexed by $x, y \in \mathbb{R}^2$ with $\|x\|_2 = \|y\|_2 = 1$. We can give a positive semidefinite factorization of size 2 of S as follows: $S_{x,y} = \langle A(x), B(y)\rangle$ where

$$A(x) = \begin{bmatrix} 1 - x_1 & x_2 \\ x_2 & 1 + x_1 \end{bmatrix} \quad B(y) = \frac{1}{2}\begin{bmatrix} 1 + y_1 & -y_2 \\ -y_2 & 1 - y_1 \end{bmatrix}.$$

Since $\|x\|_2 = \|y\|_2 = 1$ we have $A(x), B(y) \succeq 0$ and

$$\begin{aligned}\langle A(x), B(y)\rangle &= \frac{1}{2}\left[(1 - x_1)(1 + y_1) - 2x_2y_2 + (1 + x_1)(1 - y_1)\right] \\ &= 1 - \langle x, y\rangle = S_{x,y}\end{aligned}$$

as desired. Theorem 3 tells us that we must have a SDP lift of size 2, which we have already seen in Example 3.

Interpretation of Theorem 3 in terms of certificates of nonnegativity. As for the case of LP lifts, we can give an interpretation of Theorem 3 in terms of certificates of nonnegativity for valid linear inequalities.

THEOREM 4 (Gouveia, Parrilo, Thomas, [**21**]; restatement of Theorem 3). *Let C be a convex body and let X be its set of extreme points. Then C has a SDP lift of size d if, and only if, there exists a map $A : X \to \mathbf{S}^d_+$ such that the following holds: for any valid linear inequality $\ell \leq \ell_{\max}$ of C there exists $B \in \mathbf{S}^d_+$ such that*

$$\ell_{\max} - \ell(x) = \langle A(x), B\rangle \quad \forall x \in X. \tag{21}$$

Note that Equation (21) is an equality of functions on X. Just like in the LP case, it should be understood as a certificate of nonnegativity for $\ell_{\max} - \ell|_X$. Indeed the function $x \mapsto \langle A(x), B\rangle$ is "obviously" nonnegative on X since $A(x) \in \mathbf{S}^d_+$ and $B \in \mathbf{S}^d_+$.

EXAMPLE 6. In Example 4 (see also Figure 6) we saw that the square $[-1, 1]^2$ has an SDP lift of size 3. Let us exhibit the function $A : X \to \mathbf{S}^3_+$ in this case corresponding to this lift. In this example the vertex set is $X = \{-1, 1\}^2$. Consider the map A given by:

$$A(x) = \begin{bmatrix} 1 & x_1 & x_2 \\ x_1 & 1 & x_1x_2 \\ x_2 & x_1x_2 & 1 \end{bmatrix}.$$

Note that $A(x) \in \mathbf{S}^3_+$ for $x \in X$ since we have, for any $x \in X$ (using the fact that $x_1^2 = x_2^2 = 1$):

$$A(x) = \begin{bmatrix} 1 \\ x_1 \\ x_2 \end{bmatrix}\begin{bmatrix} 1 \\ x_1 \\ x_2 \end{bmatrix}^T \succeq 0. \tag{22}$$

To show that the condition of Theorem 4 is satisfied consider the facet inequality $1 - x_1 \geq 0$. Define

$$B = \frac{1}{2}\begin{bmatrix} 1 \\ -1 \\ 0 \end{bmatrix}\begin{bmatrix} 1 \\ -1 \\ 0 \end{bmatrix}^T \tag{23}$$

and note that $B \succeq 0$. Then we have for any $x \in \{-1,1\}^2$

$$\langle A(x), B\rangle = \left\langle \begin{bmatrix} 1 & x_1 & x_2 \\ x_1 & 1 & x_1x_2 \\ x_2 & x_1x_2 & 1 \end{bmatrix}, \frac{1}{2}\begin{bmatrix} 1 & -1 & 0 \\ -1 & 1 & 0 \\ 0 & 0 & 0 \end{bmatrix} \right\rangle = \frac{1}{2}(1-2x_1+1) = 1-x_1.$$

Similarly one can show that the three other facet inequalities $1+x_1 \geq 0, 1-x_2 \geq 0$ and $1+x_2 \geq 0$ can be written as $\langle A(x), B\rangle$ with the following choices of B, respectively:

$$\frac{1}{2}\begin{bmatrix}1\\1\\0\end{bmatrix}\begin{bmatrix}1\\1\\0\end{bmatrix}^T, \quad \frac{1}{2}\begin{bmatrix}1\\0\\-1\end{bmatrix}\begin{bmatrix}1\\0\\-1\end{bmatrix}^T, \quad \frac{1}{2}\begin{bmatrix}1\\0\\1\end{bmatrix}\begin{bmatrix}1\\0\\1\end{bmatrix}^T. \tag{24}$$

This shows that the map A satisfies the condition of Theorem 4.

Sums of squares. In this section we show that SDP lifts can also be interpreted in terms of *sum of squares* certificates of the facet inequalities $\ell_{\max} - \ell|_X \geq 0$. Such a certificate consists in expressing $\ell_{\max} - \ell|_X$ as a sum of squares of functions on X. More formally we have:

THEOREM 5. *Let C be a convex body and let X be its extreme points. Assume there is a subspace V of $\mathbb{R}^X$ (the space of functions $X \to \mathbb{R}$) such that the following holds:*

(*) *for any valid linear inequality $\ell \leq \ell_{\max}$ of C there are elements $h_1, \dots, h_J \in V$ such that*

$$\ell_{\max} - \ell|_X = \sum_{j=1}^{J} h_j^2. \tag{25}$$

Then C has a SDP lift of size $\dim V$.

Conversely if C has a SDP lift of size d, then there is a subspace V of $\mathbb{R}^X$ of dimension at most d^2 such that condition (*) holds.

REMARK 3. The statement above does not say anything about the structure of the functions in V (e.g., if they are polynomials, rational functions, etc.). Actually one can show with some additional work that the subspace V can be chosen to be spanned by *semialgebraic functions* [**12**]. One cannot assume however that the functions in V are all *polynomials*. In fact the case where V is the space of polynomials of bounded degree correspond to Lasserre/theta-body lifts [**20**, **29**]. In [**33**, Example 3.7] an example of a convex set with a semidefinite lift but with no Lasserre lift is given.

PROOF. We show how to prove Theorem 5 from Theorem 4. We start by proving the first part. Assume we have a subspace V of dimension d such that condition (*) holds. Let $f_1, \dots, f_d$ be a basis of this subspace and define the map $A : X \to \mathbf{S}_+^d$ as follows:

$$A(x) = \begin{bmatrix} f_1(x) \\ \vdots \\ f_d(x) \end{bmatrix}\begin{bmatrix} f_1(x) \\ \vdots \\ f_d(x) \end{bmatrix}^T = [f_i(x)f_j(x)]_{1\leq i,j\leq d}.$$

Clearly $A(x)$ is positive semidefinite (note also that $A(x)$ is rank-one for any $x \in X$). Now let $\ell \leq \ell_{\max}$ be a valid inequality for C. By (*) we know that there exists elements $h_1, \ldots, h_J \in V$ such that (25) holds. We will show that there exists $B \in \mathbf{S}_+^d$ such that $\ell_{\max} - \ell|_X = \langle A(\cdot), B\rangle$. Since each h_j is in V, we can write $h_j(x) = b_j^T f(x)$ where $b_j \in \mathbb{R}^d$ and $f(x)$ is the column vector $[f_1(x); \ldots; f_d(x)]$. Now note that $(h_j(x))^2 = (b_j^T f(x))^2 = \langle f(x)f(x)^T, b_j b_j^T\rangle = \langle A(x), b_j b_j^T\rangle$. Thus if we let $B = \sum_{j=1}^J b_j b_j^T$ we get

$$\ell_{\max} - \ell(x) = \langle A(x), B\rangle \quad \forall x \in X$$

which is what we wanted. Thus using Theorem 4, this shows that P has an SDP lift of size d.

We now prove the other direction. Assume that C has a SDP lift of size d. According to Theorem 4 this means that there exists $A : X \to \mathbf{S}_+^d$ such that the following holds: for any valid inequality $\ell \le \ell_{\max}$ there is $B \in \mathbf{S}_+^d$ such that $\ell_{\max} - \ell(x) = \langle A(x), B\rangle$, for all $x \in X$. Since $A(x) \in \mathbf{S}_+^d$ we can factorize it as $A(x) = R(x)R(x)^T$. Let V be the subspace of $\mathbb{R}^X$ spanned by the entries of R, i.e., $V = \operatorname{span}(x \mapsto R_{ij}(x), i,j = 1, \ldots, d) \subset \mathbb{R}^X$ and note that[3] $\dim V \le d^2$. Now given $\ell \le \ell_{\max}$ a valid inequality of P we know that there exists $B \in \mathbf{S}_+^d$ such that $\ell_{\max} - \ell(x) = \langle A(x), B\rangle$ for all $x \in X$. We can write B as $B = CC^T$ to get

$$\ell_{\max} - \ell(x) = \langle R(x)R(x)^T, CC^T\rangle = \|C^T R(x)\|_F^2.$$

Since each entry of $x \mapsto C^T R(x)$ is an element of V, the previous equation gives a sum-of-squares certificate of $\ell_{\max} - \ell|_X$ using functions from V. This completes the proof. □

EXAMPLE 7. The lift of the square $[-1,1]^2$ discussed earlier can be explained in terms of sum-of-squares certificates. Note that the valid inequality $1 - x_1 \ge 0$ can be certified using sum-of-squares as follows:

$$1 - x_1 = \frac{1}{2}(1 - x_1)^2 \quad \forall x \in \{-1,1\}^2. \tag{26}$$

It is crucial to note that the equality above is understood on $\{-1,1\}^2$ (the equality is of course not true globally because the left hand side is a polynomial of degree 1 whereas the right-hand side is a polynomial of degree 2). Since we are working on $\{-1,1\}^2$ the right-hand side expands to $\frac{1}{2}(1 - 2x_1 + x_1^2) = \frac{1}{2}(1 - 2x_1 + 1) = 1 - x_1$ where we used the fact that $x_1^2 = 1$. Similarly one can show that the other facet inequalities have the following sum-of-squares certificates:

$$\begin{aligned}
1 - x_2 &= \frac{1}{2}(1 - x_2)^2 \quad \forall x \in \{-1,1\}^2,\\
1 + x_1 &= \frac{1}{2}(1 + x_1)^2 \quad \forall x \in \{-1,1\}^2,\\
1 + x_2 &= \frac{1}{2}(1 + x_2)^2 \quad \forall x \in \{-1,1\}^2.
\end{aligned} \tag{27}$$

Thus if we let V be the space of polynomials of degree at most 1 on $\{-1,1\}^2$ (i.e., by an abuse of notation $V = \operatorname{span}(1, x_1, x_2)$) the condition (*) of Theorem 5 holds true. Thus this shows that $[-1,1]^2$ has a SDP lift of size $\dim V = 3$. Note that the sum-of-squares certificates in Equations (26) and (27) correspond precisely to the psd factorization given in Equations (22)–(24).

[3] As was pointed out by a reviewer, we can get the better bound $\dim V \le \binom{d+1}{2}$ if we choose $A(x) = R(x)R(x)^T$ to be a Cholesky factorization.

Proof of the factorization theorem for SDP lifts. We now give a proof of Theorem 4, which is essentially the same as the proof in [**21**, Theorem 1].

PROOF OF THEOREM 4. The proof follows the same steps as the proof of Theorem 2 for LP lifts. The only difference is that we use the generalized Farkas' lemma/strong duality for SDPs.

Assume that $C = \pi(\mathbf{S}^d_+ \cap L)$ is an SDP lift of C of size d where L is an affine subspace of $\mathbf{S}^d$ and π a linear map. Let $\ell \leq \ell_{\max}$ be a valid linear inequality for C. Since $\pi(\mathbf{S}^d_+ \cap L) \subseteq C$ the following implication holds for all $Y \in \mathbf{S}^d$:

$$\begin{cases} Y \succeq 0 \\ Y \in L \end{cases} \Rightarrow \ell_{\max} - \ell \circ \pi(Y) \geq 0.$$

We will now show using Farkas' lemma/strong duality for SDP that there is a positive semidefinite matrix $B \in \mathbf{S}^d_+$ and an affine form γ that vanishes on L such that

$$\ell_{\max} - \ell \circ \pi(Y) = \langle B, Y \rangle + \gamma(Y) \quad \forall Y \in \mathbf{S}^d. \tag{28}$$

To see why (28) is true let L_0 be the linear space in $\mathbf{S}^d$ parallel to L and let $Y_0 \in \mathbf{S}^d$ such that $L = Y_0 + L_0$. The following problems are dual to each other and the value of the primal (maximization) problem is equal to $\ell_{\max}$:

$$\begin{array}{lll} \max\limits_{Y \in \mathbf{S}^d} & (\ell \circ \pi)(Y) \\ \text{s.t.} & Y \in \mathbf{S}^d_+ \\ & Y - Y_0 \in L_0 \end{array} \qquad \begin{array}{ll} \min\limits_{B, H \in \mathbf{S}^d} & -\langle H, Y_0 \rangle \\ \text{s.t.} & -\ell \circ \pi = B + H \\ & B \in \mathbf{S}^d_+, \ H \in L_0^{\perp} \end{array} \tag{29}$$

We can assume that the intersection of L with the interior of $\mathbf{S}^d_+$ is nonempty (otherwise the intersection lies on a strict face of $\mathbf{S}^d_+$ which means that one can reduce the size of the SDP lift). In this case strong duality holds, the optimal values of the two SDPs (29) are equal to $\ell_{\max}$ and the dual (minimization) problem is attained. Let B, H be the optimal points of the dual problem in (29). From dual feasibility we have $-\ell \circ \pi = B + H$ and so since $\ell_{\max} = -\langle H, Y_0 \rangle$ we get that:

$$\ell_{\max} - \ell \circ \pi = B + H - \langle H, Y_0 \rangle.$$

Note that this shows (28) where the affine map γ is $\gamma(Y) = \langle H, Y - Y_0 \rangle$. Recall that X is the set of extreme points of C. For $x \in X$, let $A(x)$ be any element in $\mathbf{S}^d_+ \cap L$ such that $\pi(A(x)) = x$. Evaluating (28) at $A(x)$, for any $x \in X$ we get:

$$\ell_{\max} - \ell(x) = \langle B, A(x) \rangle$$

where we used the fact that $\pi(A(x)) = x$ and that $\gamma(A(x)) = 0$ since $A(x) \in L$. This proves our claim.

We now prove the converse. We show how to construct an SDP lift of P from a psd factorization of its slack matrix. By bipolarity we know that $C = \{x \in \mathbb{R}^n : \ell(x) \leq 1 \ \forall \ell \in C^\circ\}$ is an inequality description of C. Let S be the slack matrix of C and let $S_{x,\ell} = \langle A(x), B(\ell) \rangle$ be a psd factorization of S of size d, where $A(x), B(\ell) \in \mathbf{S}^d_+$ (here $x \in X$ is an extreme point of C and ℓ is an extreme point of C°). It is easy to verify that we have the following description of C:

$$C = \left\{ x \in \mathbb{R}^n : \exists A \in \mathbf{S}^d_+, \text{ s.t. } 1 - \ell(x) = \langle A, B(\ell) \rangle \ \forall \ell \in \text{extreme}(C^\circ) \right\} \tag{30}$$

To see why the inclusion "$\subseteq$" holds let $x \in X$ be an extreme point of C and take $A = A(x)$ in the right-hand side. The reverse inclusion "$\supseteq$" follows immediately by observing that $\langle A, B(\ell)\rangle \geq 0$ since A and $B(\ell)$ are positive semidefinite. It remains to show that one can put (30) into the form $C = \pi(\mathbf{S}^d_+ \cap L)$ for some linear map π and affine subspace L. This can be done in a way very similar to the proof of Yannakakis' theorem for LP lifts (see paragraph following Equation (17)). □

4. Conclusion

We presented a connection between lifted representations of convex sets and factorizations of matrices. We now mention some applications of this connection, and also briefly touch on some topics that were not covered in the chapter.

One of the main applications of the slack matrix factorization is to prove obstructions to the existence of small lifts for certain polytopes. In his original paper, Yannakakis [**40**] used his newly found connection to show that the traveling salesman polytope has no polynomial-sized symmetric linear programming lift. The more recent results [**18**] and [**35**] which we quoted in Section 1, also rely on the slack matrix factorization theorem. The connection between positive semidefinite lifts and sums of squares was in turn used in [**30**] to prove that the cut polytope has no polynomial-size semidefinite lift. It was also recently used in [**38**] to disprove a conjecture of Helton and Nie; see [**12**] for a simpler account of the proof where the connection with Theorem 5 is more explicit.

The sum of squares point of view on semidefinite lifts (Theorem 5) also provides a way to *construct* lifts (or relaxations) that are smaller than the standard Lasserre hierarchy. The key is to consider a subspace V that is not of the form $V = \{\text{polynomials of degree} \leq k\}$, but rather a "specially crafted" subspace. Such special subspaces were constructed in [**15**] and [**32**] for special examples of polytopes. For a certain family of polytopes (so-called trigonometric cyclic polytopes), this was used [**14**] to produce semidefinite lifts that are vanishingly smaller than any linear programming lift.

Whereas we only considered exact lifts of convex sets, the slack matrix factorization theorem can be extended to deal with *approximate* lifts. The setting is as follows: given two convex sets $C \subset C'$, can we find a convex set D of small LP/SDP extension complexity such that $C \subset D \subset C'$? A simple modification to the proof of Yannakakis' theorem shows that one needs to study nonnegative/psd factorizations of a certain *generalized* slack matrix for the pair (C, C'); see [**34**, Section 4.1] for details. This connection was used to prove various results on the hardness of approximation of certain convex sets or polytopes, see e.g., [**5, 7, 10**].

We have focused in this chapter on linear programming and semidefinite programming lifts. In general, given a convex cone K, we say that C has a K-lift if $C = \pi(K \cap L)$ where π is a linear map and L is an affine subspace [**21**]. Under some mild conditions on K, the existence of a K-lift is equivalent to a certain K-factorization of the slack matrix of C, see [**21**] for details. The literature on extended formulations and lifts has mostly focused on linear and semidefinite programming lifts, however there has also been some research on lifts via other cones K. For example, the paper [**16**] studies lifts via the cone of *completely positive matrices* and shows that various polytopes from combinatorial optimization admit small lifts via this cone;[4] [**22**] shows how approximate nonnegative factorizations of the slack matrix of a polytope give rise to second-order cone lifts. The paper [**11**]

[4]It is NP-hard to decide membership in the completely positive cone [**9**].

gives obstructions to the existence of second-order cone lifts; these were generalized [**1**, **36**] to more general classes of cones based on facial structure.

5. Summary

We summarize briefly the main results in this chapter.

- Let $P = \operatorname{conv}(X)$ be a polytope. The slack matrix of P is defined as $S_{x,\ell} = \ell_{\max} - \ell(x)$ where rows are indexed by extreme points $x \in X$ and columns are indexed by facets $\ell \leq \ell_{\max}$ of P. The entries of S are nonnegative.
- Finding an LP lift of P is equivalent to finding a nonnegative factorization of S (Yannakakis' theorem, Theorem 1) of the form $S = AB$ where A and B are elementwise nonnegative matrices. The size of the smallest LP lift of P is equal to the nonnegative rank of S.
- Finding an SDP lift of P is equivalent to finding a *positive semidefinite* factorization of S (Theorem 3) of the form $S_{x,\ell} = \langle A(x), B(\ell)\rangle$ where $A(x), B(\ell)$ are real symmetric $d \times d$ matrices that are positive semidefinite. The size of the smallest semidefinite programming lift is equal to the positive semidefinite rank of S.
- Sum-of-squares lifts (which are a special case of SDP lifts when the map $A(x)$ is rank-one) consist in finding a subspace V of $\mathbb{R}^X$ such that any facet inequality $\ell_{\max} - \ell|_X \geq 0$ has a certificate:

$$\ell_{\max} - \ell|_X = \sum_{j=1}^{J} h_j^2$$

  where $h_1, \dots, h_J \in V$. Such a subspace yields an SDP lift of P of size $\dim V$.

References

[1] Gennadiy Averkov, *Optimal size of linear matrix inequalities in semidefinite approaches to polynomial optimization*, SIAM J. Appl. Algebra Geom. **3** (2019), no. 1, 128–151, DOI 10.1137/18M1201342. MR3922906 ↑55

[2] Egon Balas, *Disjunctive programming*, Ann. Discrete Math. **5** (1979), 3–51. Discrete optimization (Proc. Adv. Res. Inst. Discrete Optimization and Systems Appl., Banff, Alta., 1977), II. MR558566 ↑40

[3] Aharon Ben-Tal and Arkadi Nemirovski, *On polyhedral approximations of the second-order cone*, Math. Oper. Res. **26** (2001), no. 2, 193–205, DOI 10.1287/moor.26.2.193.10561. MR1895823 ↑43

[4] G. Blekherman, P. A. Parrilo, and R. R. Thomas, *Semidefinite optimization and convex algebraic geometry*, SIAM, 2013. 41

[5] Gábor Braun, Sebastian Pokutta, Samuel Fiorini, and David Steurer, *Approximation limits of linear programs (beyond hierarchies)*, 2012 IEEE 53rd Annual Symposium on Foundations of Computer Science—FOCS 2012, IEEE Computer Soc., Los Alamitos, CA, 2012, pp. 480–489. MR3186635 ↑41, 54

[6] Mark Braverman and Ankur Moitra, *An information complexity approach to extended formulations*, STOC'13—Proceedings of the 2013 ACM Symposium on Theory of Computing, ACM, New York, 2013, pp. 161–170, DOI 10.1145/2488608.2488629. MR3210777 ↑41

[7] Siu On Chan, James R. Lee, Prasad Raghavendra, and David Steurer, *Approximate constraint satisfaction requires large LP relaxations*, 2013 IEEE 54th Annual Symposium on Foundations of Computer Science—FOCS 2013, IEEE Computer Soc., Los Alamitos, CA, 2013, pp. 350–359, DOI 10.1109/FOCS.2013.45. MR3246237 ↑41, 54

[8] Michele Conforti, Gérard Cornuéjols, and Giacomo Zambelli, *Extended formulations in combinatorial optimization*, 4OR **8** (2010), no. 1, 1–48, DOI 10.1007/s10288-010-0122-z. MR2609619 ↑40, 43

[9] Peter J. C. Dickinson and Luuk Gijben, *On the computational complexity of membership problems for the completely positive cone and its dual*, Comput. Optim. Appl. **57** (2014), no. 2, 403–415, DOI 10.1007/s10589-013-9594-z. MR3165055 ↑54

[10] H. Fawzi, *On polyhedral approximations of the positive semidefinite cone*, arXiv preprint `arXiv:1811.09649`, 2018. 54

[11] Hamza Fawzi, *On representing the positive semidefinite cone using the second-order cone*, Math. Program. **175** (2019), no. 1-2, Ser. A, 109–118, DOI 10.1007/s10107-018-1233-0. MR3942887 ↑54

[12] H. Fawzi, *The set of separable states has no finite semidefinite representation except in dimension* 3×2, arXiv preprint `arXiv:1905.02575`, 2019. 51, 54

[13] Hamza Fawzi, João Gouveia, Pablo A. Parrilo, Richard Z. Robinson, and Rekha R. Thomas, *Positive semidefinite rank*, Math. Program. **153** (2015), no. 1, Ser. B, 133–177, DOI 10.1007/s10107-015-0922-1. MR3395545 ↑49

[14] Hamza Fawzi, James Saunderson, and Pablo A. Parrilo, *Sparse sums of squares on finite abelian groups and improved semidefinite lifts*, Math. Program. **160** (2016), no. 1-2, Ser. A, 149–191, DOI 10.1007/s10107-015-0977-z. MR3555385 ↑41, 54

[15] Hamza Fawzi, James Saunderson, and Pablo A. Parrilo, *Equivariant semidefinite lifts of regular polygons*, Math. Oper. Res. **42** (2017), no. 2, 472–494, DOI 10.1287/moor.2016.0813. MR3652002 ↑54

[16] Samuel Fiorini, Serge Massar, Manas K. Patra, and Hans Raj Tiwary, *Generalized probabilistic theories and conic extensions of polytopes*, J. Phys. A **48** (2015), no. 2, 025302, 22, DOI 10.1088/1751-8113/48/2/025302. MR3293065 ↑54

[17] Samuel Fiorini, Serge Massar, Sebastian Pokutta, Hans Raj Tiwary, and Ronald de Wolf, *Linear vs. semidefinite extended formulations: exponential separation and strong lower bounds*, STOC'12—Proceedings of the 2012 ACM Symposium on Theory of Computing, ACM, New York, 2012, pp. 95–106, DOI 10.1145/2213977.2213988. MR2961499 ↑40

[18] Samuel Fiorini, Serge Massar, Sebastian Pokutta, Hans Raj Tiwary, and Ronald de Wolf, *Exponential lower bounds for polytopes in combinatorial optimization*, J. ACM **62** (2015), no. 2, Art. 17, 23, DOI 10.1145/2716307. MR3346156 ↑54

[19] Michel X. Goemans, *Smallest compact formulation for the permutahedron*, Math. Program. **153** (2015), no. 1, Ser. B, 5–11, DOI 10.1007/s10107-014-0757-1. MR3395538 ↑41, 43

[20] João Gouveia, Pablo A. Parrilo, and Rekha R. Thomas, *Theta bodies for polynomial ideals*, SIAM J. Optim. **20** (2010), no. 4, 2097–2118, DOI 10.1137/090746525. MR2630035 ↑51

[21] João Gouveia, Pablo A. Parrilo, and Rekha R. Thomas, *Lifts of convex sets and cone factorizations*, Math. Oper. Res. **38** (2013), no. 2, 248–264, DOI 10.1287/moor.1120.0575. MR3062007 ↑49, 50, 53, 54

[22] João Gouveia, Pablo A. Parrilo, and Rekha R. Thomas, *Approximate cone factorizations and lifts of polytopes*, Math. Program. **151** (2015), no. 2, Ser. B, 613–637, DOI 10.1007/s10107-014-0848-z. MR3348165 ↑54

[23] João Gouveia and Rekha Thomas, *Convex hulls of algebraic sets*, Handbook on semidefinite, conic and polynomial optimization, Internat. Ser. Oper. Res. Management Sci., vol. 166, Springer, New York, 2012, pp. 113–138, DOI 10.1007/978-1-4614-0769-0_5. MR2894693 ↑41

[24] M. Grötschel, L. Lovász, and A. Schrijver, *The ellipsoid method and its consequences in combinatorial optimization*, Combinatorica **1** (1981), no. 2, 169–197, DOI 10.1007/BF02579273. MR625550 ↑39

[25] J. William Helton and Jiawang Nie, *Sufficient and necessary conditions for semidefinite representability of convex hulls and sets*, SIAM J. Optim. **20** (2009), no. 2, 759–791, DOI 10.1137/07070526X. MR2515796 ↑41

[26] V. Kaibel, *Extended formulations in combinatorial optimization*, Arxiv preprint `arXiv:1104.1023`, Appeared in Optima 85 newsletter of the Mathematical Optimization Society, 2011. 43

[27] Volker Kaibel and Kanstantsin Pashkovich, *Constructing extended formulations from reflection relations*, Integer programming and combinatorial optimization, Lecture Notes in Comput. Sci., vol. 6655, Springer, Heidelberg, 2011, pp. 287–300, DOI 10.1007/978-3-642-20807-2_23. MR2820916 ↑41

[28] Volker Kaibel, Kanstantsin Pashkovich, and Dirk Oliver Theis, *Symmetry matters for sizes of extended formulations*, SIAM J. Discrete Math. **26** (2012), no. 3, 1361–1382, DOI 10.1137/110839813. MR3022143 ↑41

[29] Jean B. Lasserre, *Convex sets with semidefinite representation*, Math. Program. **120** (2009), no. 2, Ser. A, 457–477, DOI 10.1007/s10107-008-0222-0. MR2505746 ↑41, 51

[30] James R. Lee, Prasad Raghavendra, and David Steurer, *Lower bounds on the size of semidefinite programming relaxations*, STOC'15—Proceedings of the 2015 ACM Symposium on Theory of Computing, ACM, New York, 2015, pp. 567–576. MR3388236 ↑40, 54

[31] C. H. Lim and S. Wright, *Beyond the Birkhoff polytope: Convex relaxations for vector permutation problems*, Advances in Neural Information Processing Systems, 2014, pp 2168–2176. 42

[32] Monaldo Mastrolilli, *High degree sum of squares proofs, Bienstock-Zuckerberg hierarchy and CG cuts*, Integer programming and combinatorial optimization, Lecture Notes in Comput. Sci., vol. 10328, Springer, Cham, 2017, pp. 405–416, DOI 10.1007/978-3-319-59250-3. MR3678801 ↑54

[33] Tim Netzer, Daniel Plaumann, and Markus Schweighofer, *Exposed faces of semidefinitely representable sets*, SIAM J. Optim. **20** (2010), no. 4, 1944–1955, DOI 10.1137/090750196. MR2600247 ↑51

[34] K. Pashkovich, *Extended formulations for combinatorial polytopes*, PhD thesis, Otto-von-Guericke-Universität Magdeburg, 2012. 54

[35] Thomas Rothvoss, *The matching polytope has exponential extension complexity*, STOC'14—Proceedings of the 2014 ACM Symposium on Theory of Computing, ACM, New York, 2014, pp. 263–272, DOI 10.1145/2591796.2591834. MR3238952 ↑39, 40, 54

[36] J. Saunderson, *Limitations on the expressive power of convex cones without long chains of faces*, arXiv preprint `arXiv:1902.06401`, 2019. 55

[37] J. Saunderson, P. A. Parrilo, and A. S. Willsky, *Semidefinite descriptions of the convex hull of rotation matrices*, SIAM J. Optim. **25** (2015), no. 3, 1314–1343, DOI 10.1137/14096339X. MR3369076 ↑41

[38] Claus Scheiderer, *Spectrahedral shadows*, SIAM J. Appl. Algebra Geom. **2** (2018), no. 1, 26–44, DOI 10.1137/17M1118981. MR3755652 ↑41, 54

[39] C. Vinzant, *Geometry of spectrahedra*, Sums of Squares: Theory and Practice.,P. A. Parrilo and R. R. Thomas, editors, AMS, 2019. 47

[40] Mihalis Yannakakis, *Expressing combinatorial optimization problems by linear programs*, J. Comput. System Sci. **43** (1991), no. 3, 441–466, DOI 10.1016/0022-0000(91)90024-Y. MR1135472 ↑40, 44, 45, 54

Centre for Mathematical Sciences, Wilberforce Road, Cambridge, CB3 United Kingdom

Email address: `h.fawzi@damtp.cam.ac.uk`

Proceedings of Symposia in Applied Mathematics
Volume **77**, 2020
https://doi.org/10.1090/psapm/077/00683

Algebraic geometry and sums of squares

Mauricio Velasco

Abstract. We give a self-contained introduction to the geometry of real projective varieties. We then study the relationship between nonnegative polynomials and sums of squares on such varieties and explain a recent geometric characterization of projective varieties for which these two concepts coincide. This characterization unifies and explains several well-known theorems and connects sums of squares with one of the cornerstones of classical algebraic geometry, the classification of varieties of minimal degree.

1. Introduction

Characterizing multivariate nonnegative polynomials with real coefficients is arguably the central problem of both polynomial optimization and real algebraic geometry [**19**], [**8**]. A sufficient condition for a polynomial to be nonnegative is that it can be written as a sum of squares of polynomials. This sufficient condition is extremely useful because it is easily testable (as a semidefinite programming feasibility problem) and has led to a wealth of algorithms and applications in both real algebraic geometry and optimization (see for instance the books [**4**] and [**19**]). Unfortunately, it is not the case that every nonnegative polynomial can be written as a sum of squares of polynomials. This is well-known since the time of Hilbert, who in 1888 completely characterized the cases where equality holds:

Theorem 1.1 (Hilbert-1888 [**15**]). *Every nonnegative form (i.e. homogeneous polynomial) of degree $2d$ in n-variables is a sum of squares of forms of degree d if and only if either,*

(1) $n = 2$ *(bivariate forms) or*
(2) $d = 1$ *(quadratic forms) or*
(3) $n = 3$ *and* $d = 2$ *(ternary quartics).*

Since then, a significant amount of work has been devoted to trying to understand the relationship between nonnegative polynomials and sums of squares when either nonnegativity is only required on some subset $X \subseteq \mathbb{R}^n$ or when the support of the polynomials has a special combinatorial structure. For two important results in these two directions consider:

2010 *Mathematics Subject Classification.* Primary 14-02, 90-02.

Key words and phrases. Nonnegative polynomials, sums of squares, varieties of minimal degree.

THEOREM 1.2 (S-Lemma of Jakubovič-1971 [**16**]). *Let* $Q(x_1,\dots,x_n)$ *be an indefinite quadratic form and let* X *be the set of zeroes of* Q *in* $\mathbb{R}^n$*. Every quadratic form* F *which is nonnegative on* X *is of the form* $F=\ell_1^2+\cdots+\ell_k^2+\lambda Q$ *for some linear forms* $\ell_i(x_1,\dots,x_n)$ *and some real number* λ*.*

THEOREM 1.3 (Choi-Lam-Reznick-1980 [**9**]). *Every bi-homogeneous nonnegative polynomial of degree* $(2d_1,2d_2)$ *in two sets of variables of sizes* n_1,n_2 *is a sum of squares of bi-homogeneous forms if and only if either* $(n_1\leq 2, d_2\leq 1)$ *or* $(d_1\leq 1, n_2\leq 2)$*.*

In this Chapter we will unify all these Theorems by characterizing the relationship between the cone of nonnegative forms and the cone of sums of squares on arbitrary (totally) real projective varieties $X\subseteq\mathbb{P}^n$. This point of view will allow us to give a common geometric explanation to all known cases of equality between nonnegative forms and sums of squares. Moreover, the language of algebraic geometry will give us a context in which this relationship can be characterized in terms that are both simpler and more general. More concretely, the aim of this Chapter is to explain, in a self-contained manner, the following Theorem [**6**],

THEOREM 1.4. *Let* $X\subset\mathbb{P}^n$ *be a real, irreducible and non-degenerate projective variety such that the set* $X(\mathbb{R})$ *of real points is Zariski dense in* X*. Every nonnegative real quadratic form on* X *is a sum of squares of linear forms in* $\mathbb{R}[X]$ *if and only if* X *is a variety of minimal degree (i.e. one for which the equality* $\deg(X)=\operatorname{codim}(X)+1$ *holds).*

The material in this Chapter is organized as follows: Section 2 will be an introduction to projective algebraic geometry over the reals. We will introduce the basic language and the two key concepts of *dimension* and *degree.* Section 3 will consist of a proof of Theorem 1.4 and a discussion of the classification of varieties of minimal degree. Several exercises and useful references are interspersed throughout the text.

2. Projective geometry and sums of squares.

2.1. Background and conventions. In this Section we establish the notation and basic conventions about algebraic varieties used throughout the Chapter. After reading this Section the reader should be able to make sense of the sentence "Let $X\subset\mathbb{P}^n$ be a real and non-degenerate projective variety such that the set $X(\mathbb{R})$ of real points is Zariski dense" and to understand the concepts of dimension and degree. Although the treatment is logically self-contained it may be too brief to be sufficiently informative if this is your first exposure to varieties. I have therefore included references for more complete texts. The reader familiar with the basic language of varieties should skip ahead to Section 2.3 for some key examples of projective varieties or to Section 2.4 where the material on sums of squares on varieties begins.

2.1.1. *Projective space.* For a positive integer n let $\mathbb{P}^n$ be the projective n-space over the complex numbers, that is the space of lines in $\mathbb{C}^{n+1}$. More precisely, as a set $\mathbb{P}^n$ is the collection of equivalence classes in $\mathbb{C}^{n+1}\setminus\{0\}$ of the relation $u\sim v$ if and only if there exists $\lambda\in\mathbb{C}^*:=\mathbb{C}\setminus\{0\}$ such that $u=\lambda v$. We will denote points in $\mathbb{P}^n$ as $[\alpha]$ for $\alpha\in\mathbb{C}^{n+1}\setminus\{0\}$ and write them as $(n+1)$-tuples $[\alpha_0:\dots:\alpha_n]$ called the *homogeneous coordinates* of the point. We will call any vector $u\in\mathbb{C}^{n+1}\setminus\{0\}$

with $[u] = [\alpha]$ an *affine representative* for $[\alpha]$. Note that a given point has several affine representatives and therefore several sets of homogeneous coordinates whose components differ by multiplication by a common scalar. For further introductory background on $\mathbb{P}^n$ refer to [**10**, Chapter 8].

Let $\mathbb{C}[\mathbb{P}^n] := \mathbb{C}[X_0, \dots, X_n]$ be the set of polynomials with complex coefficients in the variables $X_0, \dots, X_n$. An element $F \in \mathbb{C}[\mathbb{P}^n]$ is *homogeneous of degree* d (also known as a form of degree d) if it is a sum of monomials of degree d in $X_0, \dots, X_n$ or equivalently if the equality $F(tX_0, \dots, tX_n) = t^d F(X_0, \dots, X_n)$ holds for every $t \in \mathbb{C}$. If F is a homogeneous polynomial and $[\alpha] = [\beta]$ in $\mathbb{P}^n$ then $\alpha = \lambda\beta$ and therefore $F(\alpha) = \lambda^d F(\beta)$. It follows that the set $V(F)$ given by those $[\alpha] \in \mathbb{P}^n$ for which $F(\alpha) = 0$ is a well-defined subset of $\mathbb{P}^n$.

An ideal $I \subseteq \mathbb{C}[\mathbb{P}^n]$ is a subset closed under addition and closed under multiplication by elements of $\mathbb{C}[\mathbb{P}^n]$. If $\{g_1, \dots, g_m\} \subseteq \mathbb{C}[\mathbb{P}^n]$ then the ideal generated by $g_1, \dots, g_m$ (i.e. the smallest ideal containing the g_i) is given by

$$(g_1, \dots, g_m) := \left\{ \sum_{i=1}^{m} r_i g_i : r_i \in \mathbb{C}[\mathbb{P}^n] \right\}.$$

A fundamental result, the Hilbert basis Theorem [**1**, Theorem 7.5] says that every ideal $I \subseteq \mathbb{C}[\mathbb{P}^n]$ is of the form $(g_1, \dots, g_m)$ for some *finite* set $g_1, \dots, g_m \in \mathbb{C}[\mathbb{P}^n]$ of generators. An ideal $I \subseteq \mathbb{C}[\mathbb{P}^n]$ is homogeneous if it has a set of homogeneous generators. For further background on ideals in polynomial rings see [**10**, Section 1.4]. If $S \subseteq \mathbb{C}[\mathbb{P}^n]$ is any homogeneous ideal then the *algebraic set* $V(S)$ defined by S is given by

$$V(S) := \{[\alpha] \in \mathbb{P}^n : F(\alpha) = 0 \text{ for every homog. polynomial } F \text{ in } S\}.$$

In words, the algebraic subsets of $\mathbb{P}^n$ are the solutions of systems of homogeneous polynomial equations.

EXERCISE 2.1. Show that every algebraic set $V(S)$ is of the form $V(g_1) \cap \dots \cap V(g_k)$ for some collection of finitely many homogeneous polynomials $g_1, \dots, g_k \in \mathbb{C}[\mathbb{P}^n]$ (i.e. the set of common solutions of finitely many homogeneous polynomial equations).

The collection of all algebraic subsets of $\mathbb{P}^n$ is closed under arbitrary intersections and finite unions and contains both $\mathbb{P}^n = V(\{0\})$ and $\emptyset = V((1))$. The algebraic sets are therefore the closed sets of a topology on $\mathbb{P}^n$ called the *Zariski topology.* Throughout this Chapter we will think of $\mathbb{P}^n$ as a topological space with the Zariski topology. An algebraic set Z is called *reducible* if there exists two different $Z_1, Z_2 \subseteq Z$, proper closed subsets of Z, such that $Z = Z_1 \cup Z_2$ and is called *irreducible* otherwise.

EXERCISE 2.2. Use the Hilbert basis Theorem to prove that every closed set $Z \subseteq \mathbb{P}^n$ can be written uniquely as an irredundant union of irreducible closed sets called the irreducible components of Z.

EXERCISE 2.3. Prove that $X = V(x^2 - xw + y^2 + yz, xy + zw) \subseteq \mathbb{P}^3$ is reducible and find its irreducible components.

2.1.2. *Projective varieties.* A *projective subvariety* of $\mathbb{P}^n$ is an algebraic set $X \subseteq \mathbb{P}^n$ endowed with the Zariski topology (i.e. as a topological subspace of $\mathbb{P}^n$). In this section we will assign a commutative ring to each projective subvariety $X \subseteq$

$\mathbb{P}^n$, its *homogeneous coordinate ring* $\mathbb{C}[X]$. As you will see in our treatment of dimension and degree the algebraic object $\mathbb{C}[X]$ captures many geometric properties of the set X in a manner that makes them more accessible and often effectively computable. The construction will require the use of quotient rings so we begin with some preliminaries. Finally, we will discuss the (restricted) concept of morphism of projective varieties which we will use in this Chapter.

Quotient rings. If $I \subseteq \mathbb{C}[\mathbb{P}^n]$ is an ideal then we can define a *quotient ring* $R := \mathbb{C}[\mathbb{P}^n]/I$ as the largest ring where the elements of I become zero. More precisely R is the set of equivalence classes of the equivalence relation $\sim$ in $\mathbb{C}[\mathbb{P}^n]$ given by $p \sim q$ if and only $p - q \in I$. We denote the equivalence class of p by $p + I$ and define the sum and multiplication of classes by $(a+I)+(b+I) := a+b+I$ and $(a+I)(b+I) := ab+I$ making R into a commutative ring. For further background on quotient rings see [**10**, Section 5.2]. The following very useful exercise explains how to construct ring homomorphisms from quotient rings,

EXERCISE 2.4. Let S be any commutative ring. Prove that there is a bijective correspondence between the ring homomorphisms $\phi : \mathbb{C}[\mathbb{P}^n] \to S$ which map the generators of I to zero and the ring homomorphisms $R \to S$.

If I is an ideal generated by homogeneous elements then the quotient ring R inherits the structure of a *graded ring* from $\mathbb{C}[\mathbb{P}^n]$ in the sense that:

(1) For every integer d the classes in R of homogeneous polynomials in $\mathbb{C}[\mathbb{P}^n]$ of degree d together with zero form a vector subspace $R_d \subseteq R$.
(2) Any two distinct subspaces R_d intersect only at zero.
(3) The decomposition is compatible with multiplication in the sense that $R_a R_b \subseteq R_{a+b}$ for all $a, b \in \mathbb{Z}$.

Note that (1) and (2) imply that, as a vector space $R = \bigoplus_{d\in\mathbb{Z}} R_d$.

Homogeneous coordinate rings. The *ideal of definition* $I(X)$ of a subvariety $X \subseteq \mathbb{P}^n$ is given by

$$I(X) := (\{F \in \mathbb{C}[\mathbb{P}^n] : \ F \text{ homog. with } F(\alpha) = 0 \text{ for every } [\alpha] \in X\})$$

The *homogeneous coordinate ring of* X is the graded ring defined as $\mathbb{C}[X] := \mathbb{C}[\mathbb{P}^n]/I(X)$.

Regular maps between projective varieties. If $X \subseteq \mathbb{P}^n$ and $Y \subseteq \mathbb{P}^m$ are projective varieties and $F_0, \ldots, F_m \in \mathbb{C}[X]$ are polynomials of the same degree with no common zero on X then we can use them to define a function $\phi(\alpha_0 : \cdots : \alpha_n) := [F_0(\alpha) : \cdots : F_m(\alpha)]$ which maps X to $\mathbb{P}^m$. If $\phi(X) \subseteq Y$ then ϕ is a *regular map* between X and Y. It is known that every regular map between projective varieties is continuous and Zariski closed (i.e. maps closed sets to closed sets).

EXAMPLE 2.5. If $A \in \mathbb{C}^{(n+1)\times(n+1)}$ is an invertible linear map and $L_0, \ldots, L_n$ are the linear forms given by the components of the product $A(X_0, \ldots, X_n)^t$ then the map $\phi_A(X_0 : \cdots : X_n) = (L_0 : \cdots : L_n)$ defines an automorphism of $\mathbb{P}^n$. Two invertible matrices $A_1, A_2 \in GL(n+1)$ define the same automorphism if and only if they differ by multiplication by a nonzero scalar. It is known that every automorphism of $\mathbb{P}^n$ is of this form (so $\mathrm{Aut}(\mathbb{P}^n) = PGL(n+1) := GL(n+1)/\mathbb{C}^*I$).

REMARK 2.6. The curious reader may wonder why we are introducing the term projective variety if a projective variety is just an algebraic set. The answer is that a projective variety X is an algebraic set together with a local concept of *regular functions* on open sets of X (in the same way that a differentiable manifold is a topological space together with a concept of differentiable function on open sets). This local structure allows us to define morphisms between varieties and therefore is very important for the development of the general theory. In this Chapter we will only use regular maps of a very special kind (those described in the previous paragraph) and our incomplete definition of projective variety and regular map will (mostly) suffice. The reader who wishes to understand what varieties really are is encouraged to read Section 1 of the excellent textbook [**18**].

2.2. Real projective varieties. We say that a projective variety $X \subseteq \mathbb{P}^n$ is *defined over* $\mathbb{R}$ or that X is a *real projective variety* if the ideal $I(X)$ admits a set of generators, as an ideal in $\mathbb{C}[\mathbb{P}^n]$, which are polynomials with real coefficients. In that case we define the *(real) homogeneous coordinate ring* of X as

$$\mathbb{R}[X] := \mathbb{R}[X_0, \ldots, X_n]/(I(X) \cap \mathbb{R}[X_0, \ldots, X_n]).$$

If X is a real projective variety and $k \in \{\mathbb{R}, \mathbb{C}\}$ we write $X(k)$ to denote the subset of points of X which have some affine representative with components in k. A regular map $\phi : X \to Y$ is *defined over* $\mathbb{R}$ whenever it has components $F_0, \ldots, F_m \in \mathbb{R}[X]$. We say that X is *totally real* if $X(\mathbb{R})$ is Zariski dense in X.

EXAMPLE 2.7. (Plane conics) If $X \subseteq \mathbb{P}^2$ is the quadratic hypersurface $V(X_0^2 + X_1^2 + X_2^2)$ then X is defined over $\mathbb{R}$ and $X(\mathbb{R}) = \emptyset$. If $Y \subseteq \mathbb{P}^2$ is given by $V(Y_0Y_1 - Y_2^2)$ then Y is defined over $\mathbb{R}$ and $Y(\mathbb{R})$ is Zariski-dense in Y. The map $\phi : \mathbb{P}^2 \to \mathbb{P}^2$ given by $\phi([X_0 : X_1 : X_2]) = [X_0 + iX_1 : X_0 - iX_1 : iX_2]$ is an isomorphism between X and Y defined over $\mathbb{C}$. However, no isomorphism between X and Y defined over $\mathbb{R}$ exists since otherwise the sets of real points of X and Y would be in bijection.

2.3. Some key examples of projective varieties. In this section we give three families of examples of projective varieties: hypersurfaces, the Veronese embeddings and the Segre embeddings. The aim of the Section is to give you a more concrete understanding of what projective varieties and their coordinate rings are. We will heavily use these examples throughout the Chapter.

2.3.1. *Hypersurfaces.* Let $X \subseteq \mathbb{P}^n$ be an algebraic variety defined by a single homogeneous polynomial $F \in \mathbb{C}[\mathbb{P}^n]$. If the polynomial is irreducible (i.e. if it cannot be written as $F = GH$ where G, H are both non-constant) then X is irreducible. The homogeneous coordinate ring of X is given by $\mathbb{C}[X] := \mathbb{C}[\mathbb{P}^n]/(F)$.

EXAMPLE 2.8. (Hyperplanes) If $X \subseteq \mathbb{P}^n$ is the subvariety defined by a non-zero linear form $L(X_0, \ldots, X_n) \in \mathbb{C}[\mathbb{P}^n]$ then X is irreducible and is called a hyperplane. Its coordinate ring is $\mathbb{C}[X] = \mathbb{C}[\mathbb{P}^n]/(L)$. If $L = X_n + \sum_{i=0}^{n-1} c_i X_i$ then $\mathbb{C}[X] \cong \mathbb{C}[X_0, \ldots, X_{n-1}] \cong \mathbb{C}[\mathbb{P}^{n-1}]$ and therefore $X \cong \mathbb{P}^{n-1}$ with isomorphism given by $[X_0 : \cdots : X_{n-1}] \to [X_0 : \cdots : X_{n-1} : -\sum_{i=0}^{n-1} c_i X_i]$.

EXAMPLE 2.9. (A plane conic) If $X \subseteq \mathbb{P}^2$ is given by $V(X_0X_1 - X_2^2)$ then

$$\mathbb{C}[X] := \mathbb{C}[X_0, X_1, X_2]/(X_0X_1 - X_2^2)$$

and in particular $X_0X_1 = X_2^2$ in $\mathbb{C}[X]$ so the ring is not a unique factorization domain. It follows that $\mathbb{C}[X]$ is not isomorphic to any polynomial ring. Nevertheless X is isomorphic to $\mathbb{P}^1$ via the map $\phi : \mathbb{P}^1 \to X$ given by $\phi([s : t]) = [s^2 : t^2 : st]$.

EXAMPLE 2.10. (Quadric hypersurfaces) Let $Q(X_0, \ldots, X_n)$ be a non-zero quadratic form in $\mathbb{C}[\mathbb{P}^n]$. Up to the action of $\operatorname{Aut}(\mathbb{P}^n)$ the hypersurface defined by Q is completely determined by its rank k (meaning the rank of the unique complex symmetric matrix S with $\vec{x}^t S\vec{x} = Q$ where $\vec{x} = (X_0, \dots, X_n)^t$). More precisely, what this means is that Q has rank k if and only if there exists an invertible matrix $A \in GL_{\mathbb{C}}(n+1)$ defining an automorphism ϕ_A of $\mathbb{P}^n$ (see Example 2.5) such that $\phi_A(X)$ is the variety defined by $X_0^2 + \cdots + X_k^2$.

EXERCISE 2.11. *(Real vs totally real quadric hypersurfaces)*

(1) Let $Q(X_0, \dots, X_n)$ be a real quadric. Prove that $X = V(Q)$ is totally-real if and only if Q has at least one positive and one negative eigenvalue (Hint: use the fact that an irreducible real variety X is totally real if and only if it contains a smooth real point).
(2) Construct an example of a quadratic hypersurface $X \subseteq \mathbb{P}^3$ defined over $\mathbb{R}$ for which $X(\mathbb{R})$ is a curve.

2.3.2. *The Veronese embeddings of* $\mathbb{P}^n$. Given integers d and n let $k := \binom{n+d}{d} - 1$ and let $E \subseteq \mathbb{N}^n$ be the set of exponent vectors of monomials of degree d in the variables $X_0, \dots, X_n$ in some fixed order. For $I \in E$ we let $m_I := \prod_{j=0}^n X_j^{I_j}$. Define the d-uple map $\nu_d : \mathbb{P}^n \to \mathbb{P}^k$ by the formula

$$\nu_d(\alpha_0 : \cdots : \alpha_n) = [m_I(\alpha)]_{I \in E}$$

Note that ν_d is a regular map since all monomials have the same degree and have no common zero in $\mathbb{P}^n$. In particular ν_d maps closed sets to closed sets and the *d-th Veronese embedding* of $\mathbb{P}^n$ is the projective variety $X := \nu_d(\mathbb{P}^n)$. It is known that the Veronese map is an isomorphism between $\mathbb{P}^n$ and X. If we endow $\mathbb{P}^k$ with coordinates Y_I indexed by $I \in E$ then the ideal of definition of X is given by

$$I(X) := (\{Y_I Y_J - Y_K Y_L : I + J = K + L\})$$

and in particular we see that X is a real projective variety.

EXERCISE 2.12. Let $X = \nu_d(\mathbb{P}^n)$ and let $R := \mathbb{R}[X]$. Prove that R_j is isomorphic, as a vector space, to the space of homogeneous polynomials of degree dj in $X_0, \dots, X_n$ for any natural number j. Use this to find a formula for $\dim_{\mathbb{R}}(R_j)$.

REMARK 2.13. The previous exercise explains why the Veronese variety X is extremely useful. It allows us to linearize questions about forms of degree d in $\mathbb{P}^n$ in the sense that ν_d maps the zeroes of a form of degree d in $\mathbb{P}^n$ to the intersection of X and a hyperplane isomorphically.

EXAMPLE 2.14. (Rational normal curves) An interesting case of the Veronese varieties is obtained when $n = 1$. In this case $k = d$ and $\nu_d(\mathbb{P}^1)$ is called a *rational normal curve*. In coordinates $Y_0, \dots, Y_d$ of $\mathbb{P}^d$ the ideal of definition $I(X)$ is given by

$$I(X) := \left(2 \times 2 \text{ minors of } \begin{pmatrix} Y_0 & Y_1 & \dots & Y_{d-1} \\ Y_1 & Y_2 & \dots & Y_d \end{pmatrix} \right).$$

EXERCISE 2.15. Let J be the ideal of 2×2-minors of the previous example.

(1) Prove that for every integer d the vector space spanned by the classes of monomials of degree d in $\mathbb{C}[\mathbb{P}^n]/J$ is generated by the $nd + 1$ classes of monomials of the form $Y_0^a Y_n^b Y_j^c$ where $0 < j < n$, $c \in \{0, 1\}$ and $a + b + c = d$.
(2) Conclude that the ideal of definition of the rational normal curve equals J.

EXAMPLE 2.16. (The Veronese surface) Another important special case occurs when $n = 2$ and $d = 2$. In this case $k = 5$ and $X = \nu_2(\mathbb{P}^2)$ is the *Veronese surface* in $\mathbb{P}^5$ defined as the image of the map $[X_0^2 : X_0X_1 : X_0X_2 : X_1^2 : X_1X_2 : X_2^2]$. In coordinates $Y_0, \ldots, Y_5$ its ideal of definition is given by

$$I(X) = \left(2 \times 2 \text{ minors of } \begin{pmatrix} Y_0 & Y_1 & Y_2 \\ Y_1 & Y_3 & Y_4 \\ Y_2 & Y_4 & Y_5 \end{pmatrix} \right).$$

so X is the locus of symmetric 3×3 matrices of rank one.

EXERCISE 2.17. Verify the above description of the ideal of the Veronese surface in $\mathbb{P}^5$.

2.3.3. *The Segre embedding of a product of projective spaces.* Given integers n_1, n_2 let $k = (n_1 + 1)(n_2 + 1) - 1$ and consider the map $\sigma : \mathbb{P}^{n_1} \times \mathbb{P}^{n_2} \to \mathbb{P}^k$ given by

$$\sigma([\alpha_0 : \cdots : \alpha_{n_1}], [\beta_0 : \cdots : \beta_{n_2}]) = [\alpha_i \beta_j]_{i=0,\ldots,n_1, j=0,\ldots,n_2}$$

The map σ is injective and defines an embedding of $\mathbb{P}^{n_1} \times \mathbb{P}^{n_2}$ in $\mathbb{P}^k$ called the Segre embedding of $\mathbb{P}^{n_1} \times \mathbb{P}^{n_2}$ which we denote $X = \sigma(\mathbb{P}^{n_1} \times \mathbb{P}^{n_2})$. If $\mathbb{P}^k$ is endowed with coordinates Y_{ij} with $i = 0, \ldots, n_1$ and $j = 0, \ldots, n_2$ then $I(X)$ is given by

$$I(X) := \left(2 \times 2 \text{ minors} \begin{pmatrix} Y_{00} & Y_{01} & \ldots & Y_{0n_1} \\ \vdots & \vdots & \ldots & \vdots \\ Y_{n_2 0} & Y_{n_2 1} & \ldots & Y_{n_2 n_1} \end{pmatrix} \right).$$

so X is the locus of matrices of size $(n_2 + 1) \times (n_1 + 1)$ of rank one.

REMARK 2.18. The Segre embedding X of $\mathbb{P}^{n_1} \times \mathbb{P}^{n_2}$ is important because of two reasons. First it endows the product $\mathbb{P}^{n_1} \times \mathbb{P}^{n_2}$ with the structure of a projective variety and second, it allows us to linearize questions about homogeneous multiforms of degree $(1, 1)$ (i.e. polynomials $F(X_0, \ldots, X_{n_1}, Z_0, \ldots, Z_{n_2})$ which are sums of monomials of degree one in the $X's$ and in the $Z's$). The zeroes in $\mathbb{P}^{n_1} \times \mathbb{P}^{n_2}$ of such a multiform are mapped isomorphically by σ to the intersections of X and a hyperplane.

EXAMPLE 2.19. (Full rank quadric hypersurfaces in $\mathbb{P}^3$) If $n_1 = n_2 = 1$ then $X := \sigma(\mathbb{P}^1 \times \mathbb{P}^1)$ is a quadric hypersurface in $\mathbb{P}^3$. It follows from Example 2.10 that every quadric hypersurface of full rank in $\mathbb{P}^3$ is isomorphic to X.

2.4. Nonnegative polynomials and sums of squares on projective varieties. With the basic language of real projective varieties at our disposal we can now introduce the two central objects of this Chapter, namely the cone of nonnegative quadratic forms and the cone of sums of squares of linear forms on a variety. Let $X \subseteq \mathbb{P}^n$ be a projective variety defined over $\mathbb{R}$ and let $\mathbb{R}[X]_k$ be the space of forms of degree k in $\mathbb{P}^n$ restricted to X.

DEFINITION 2.20. The cone P_X of nonnegative quadratic forms on X is given by

$$P_X := \{F \in \mathbb{R}[X]_2 : \forall [\alpha] \in X(\mathbb{R}) \, (F(\alpha) \geq 0)\}.$$

The cone Σ_X of sums of squares of linear forms on X is given by

$$\Sigma_X := \left\{ F \in \mathbb{R}[X]_2 : \exists m \in \mathbb{N}, \ell_1, \ldots, \ell_m \in \mathbb{R}[X]_1 \text{ with } F = \sum \ell_i^2 \right\}.$$

Note that a homogeneous polynomial *of even degree* has a well defined sign (positive, negative or zero) at a point $[\alpha] \in X(\mathbb{R})$ given by the sign of $F(\alpha)$ for any real affine representative of $[\alpha]$ and therefore P_X is well defined. It is immediate that Σ_X and P_X are convex cones in the real vector space $\mathbb{R}[X]_2$ and that $\Sigma_X \subseteq P_X$. The cone P_X is fundamental in optimization (see the next Section) while the cone Σ_X is *semidefinitely representable* and therefore amenable to efficient optimization. The aim of this Chapter is to understand when does the best-case scenario, namely the equality $\Sigma_X = P_X$, occur.

EXERCISE 2.21. Let $X \subseteq \mathbb{P}^n$ be any real projective variety. Prove that the cone Σ_X is a linear projection of the cone of positive semidefinite symmetric matrices of some size.

REMARK 2.22. At first glance it may seem as if, by restricting only to quadratic forms, we are looking at a rather special case of the relationship between nonnegative polynomials and sums of squares. However this is not the case because forms of degree $2d$ on a variety $Y \subseteq \mathbb{P}^m$ are in bijective correspondence with quadratic forms on the variety $X := \nu_d(Y)$ obtained by re-embedding Y via the d-th Veronese map of Section 2.3.2. In particular, every form of degree $2d$ on Y is a sum of squares of forms of degree d if and only if the analogous statement holds for quadratic forms on X.

REMARK 2.23. At this point a reader mainly interested in optimization may ask: why are we considering projective varieties at all if polynomial optimization is mainly about algebraic subsets of $\mathbb{R}^n$? Our answer to this question consists of two parts: First, because we can, in the sense that, as we show in Section 2.6 any such optimization problem can be reduced to one about nonnegativity of quadratic forms on some projective variety. And second, because we believe we should, in the sense that, as we we will see in the next Sections, projective geometry gives us a wealth of new tools (degrees, hyperplane sections, projections away from points) which allow us to clarify the relationship between nonnegative polynomials and sums of squares. We hope to convince the reader that the projective approach to sums of squares is substantially simpler and well worth of the (admittedly significant) initial effort of establishing the correct language.

2.4.1. *The basic invariants of projective varieties.* The two main numerical invariants of a projective variety $X \subseteq \mathbb{P}^n$ are its *dimension* and its *degree*. Informally, a variety has dimension at least d if the intersection of X with any set of d hyperplanes is nonempty. The dimension of X is defined as the largest integer d with this property. It turns out that the intersection of a d dimensional variety with d "random" hyperplanes is a finite set and that the cardinality of this set is independent of the chosen hyperplanes whenever these are sufficiently generic. This cardinality is precisely the definition of the term *degree* of X.

EXAMPLE 2.24. (dimension and degree of twisted cubic) Let $X \subseteq \mathbb{P}^3$ be the twisted cubic and let $\nu_3 : \mathbb{P}^1 \to \mathbb{P}^3$ be the Veronese map given by $\nu[s : t] = [s^3 : s^2t : st^2 : t^3]$. If $H = a_0X_0 + a_1X_1 + a_2X_2 + a_3X_3$ then $V(H) \cap X$ has the same cardinality as the solutions in $\mathbb{P}^1$ to the cubic equation $a_0s^3 + a_1s^2t + a_2st^2 + a_3t^3$. If $a_0, \dots, a_3$ are such that the discriminant of the cubic does not vanish (i.e. for all $a_0, \dots, a_3$ in an open set in the projective space $(\mathbb{P}^3)^*$ of hyperplanes in $\mathbb{P}^3$) then this set has cardinality three. It follows that a second randomly chosen hyperplane

H_2 will not contain any point of $X \cap V(H)$. We conclude that X has dimension one and degree three.

In some cases it may be difficult to compute dimension and degree by intersecting a variety with generic subspaces as in the previous example (it is difficult to produce explicit examples of "generic" hyperplanes, most explicit choices will involve small coefficients and thus be likely to satisfy additional equations). For this reason our formal definitions of dimension and degree will use a different, more algebraic route. We will define dimension and degree in terms of coefficients of the Hilbert Polynomial, a key numerical invariant of the homogeneous coordinate ring $\mathbb{C}[X]$ and later prove that dimension and degree have the geometric interpretations given in the first paragraph of this section. In particular we will clarify the meaning of the expression "sufficiently random" hyperplanes.

REMARK 2.25. A numerical quantity is a *numerical invariant* of some category if its value is the same on isomorphic objects of that category. For instance, if X and Y are isomorphic varieties (with the morphisms defined in Section 2.1.2 or more generally as locally ringed spaces as in [**18**]) then their dimension is the same and thus dimension is an invariant of varieties. On the other hand degree is *not* an invariant of varieties. For instance a line and the twisted cubic are isomorphic varieties in $\mathbb{P}^3$ (since both are isomorphic to $\mathbb{P}^1$) but their degrees are one and three respectively. When working with projective varieties embedded in a fixed projective space $\mathbb{P}^n$ however, one is often interested in classifying them up to *projective equivalence*, that is considering X and Y equivalent if and only there is an automorphism ϕ of $\mathbb{P}^n$ (see Example 2.5) with $\phi(X) = Y$. Under this notion of equivalence degree is a numerical invariant.

2.4.2. *Hilbert functions and polynomials.* Any polynomial in the ring $\mathbb{R}[\mathbb{P}^n]$ (or $\mathbb{C}[\mathbb{P}^n]$) can be written uniquely (by grouping monomials of the same degree) as a sum of homogeneous polynomials. More precisely, if $\mathbb{R}[\mathbb{P}^n]_j$ denotes the vector space of homogeneous polynomials in $\mathbb{R}[\mathbb{P}^n]$ of degree j together with zero then we have the following direct sum decomposition of vector spaces

$$\mathbb{R}[\mathbb{P}^n] = \bigoplus_{j \in \mathbb{N}} \mathbb{R}[\mathbb{P}^n]_j.$$

Recall that a similar decomposition, called a grading, is possible in the homogeneous coordinate ring of every projective variety.

DEFINITION 2.26. A ring R containing $\mathbb{R}$ is *positively graded* if for $j \in \mathbb{N}$ there exist vector subspaces $R_j \subseteq R$ such that:

(1) The ring R decomposes (as an $\mathbb{R}$-vector space) as $R := \bigoplus_{j \in \mathbb{N}} R_j$
(2) For all $s, t \in \mathbb{N}$ the inclusion $R_s R_t \subseteq R_{s+t}$ holds.

EXAMPLE 2.27. Let $I \subseteq \mathbb{R}[\mathbb{P}^n]$ be an ideal generated by homogeneous elements. For a natural number j define $I_j := I \cap \mathbb{R}[\mathbb{P}^n]_j$. Let $R := \mathbb{R}[\mathbb{P}^n]/I$ and define $R_j := \mathbb{R}[\mathbb{P}^n]_j / I_j$. The restriction of the quotient map $q : \mathbb{R}[\mathbb{P}^n]_j \to R$ induces an injection $q : R_j \to R$ and, abusing notation, we denote its image by $R_j \subseteq R$. This decomposition makes R into a graded ring.

If $X \subseteq \mathbb{P}^n$ is a projective variety then $I(X)$ is an ideal generated by homogeneous elements. As in the previous example, we will henceforth think of $\mathbb{R}[X]$ as a positively graded ring.

DEFINITION 2.28. If $R \supseteq \mathbb{R}$ is a graded ring we define its *Hilbert function* $\mathrm{HF}_R : \mathbb{N} \to \mathbb{N}$ by $\mathrm{HF}_R(t) := \dim_{\mathbb{R}}(R_t)$. If $X \subseteq \mathbb{P}^n$ is a projective variety then the Hilbert function of X is the Hilbert function of its homogeneous coordinate ring, given by $\mathrm{HF}_X(t) := \dim_{\mathbb{R}}(\mathbb{R}[X]_t)$.

REMARK 2.29. The Hilbert function is independent of the field over which we do our calculations. By this I mean that the Hilbert functions of $\mathbb{C}[X]$, as a graded $\mathbb{C}$-vector space and that of $\mathbb{R}[X]$ as graded $\mathbb{R}$-vector space are equal.

One of Hilbert's very significant contributions to commutative algebra is the discovery that the values of the Hilbert function agree, for all sufficiently large t, with those of a (obviously unique) polynomial. This is a consequence of his remarkable Syzygy theorem [**11**, Section 1.10].

DEFINITION 2.30. The *Hilbert Polynomial* of a graded ring $R \supseteq \mathbb{R}$ is the unique polynomial $p_R(t)$ with the property that $HF_R(t) = p_R(t)$ for all sufficiently large t. The Hilbert polynomial $p_X(t)$ of a variety $X \subseteq \mathbb{P}^n$ is the Hilbert polynomial of $\mathbb{R}[X]$.

2.4.3. *Dimension and degree.* Since the Hilbert polynomial of $X \subseteq \mathbb{P}^n$ depends only on its homogeneous coordinate ring, the coefficients of the Hilbert function give us numerical invariants of projective varieties (see Remark 2.25).

DEFINITION 2.31. Let $X \subseteq \mathbb{P}^n$ be a projective variety. If $p_X(t) = a_d t^d + o(t^{d-1})$ then we define the *dimension* of X by $\dim(X) := d$ and the *degree* of X by $\deg(X) := d!a_d$.

EXAMPLE 2.32. (Dimension and degree of twisted cubic via Hilbert polynomial) If $X \subseteq \mathbb{P}^3$ is the twisted cubic from Example 2.24 then there is an isomorphism between $\mathbb{R}[X]_t$ and the set of homogeneous polynomials of degree $3t$ in X_0, X_1 and therefore $\mathrm{HF}_X(t) = 3t + 1 = p_X(t)$. It follows that $\dim(X) = 1$ and $\deg(X) = 3$.

EXERCISE 2.33. (Some Hilbert polynomial computations)

(1) Prove that the dimension of the space of homogeneous polynomials of degree t in $n + 1$ variables is given by $\binom{n+t}{t}$.
(2) Let $F \in \mathbb{C}[\mathbb{P}^n]$ be a homogeneous polynomial of degree d and let $X = V(F)$. Compute the leading coefficient of the Hilbert polynomial of $\mathbb{C}[X]$. Conclude that $X = V(F)$ is a variety of dimension $n - 1$ and degree d.
(3) Let $X := \nu_d(\mathbb{P}^n)$ be the d-th Veronese embedding of $\mathbb{P}^n$. Find the leading term of the Hilbert polynomial of X and determine the degree and dimension of X. (Exercise 2.12 could be useful here).

The following Lemma describes the Hilbert function and Hilbert polynomial of a finite set $X \subseteq \mathbb{P}^n$.

LEMMA 2.34. *Let $X \subseteq \mathbb{P}^n$ be a finite set of cardinality δ. The following statements hold:*

(1) *The Hilbert function of X satisfies $HF_X(0) = 1$ and is strictly increasing until it attains the value δ, after which it becomes constant.*
(2) *The dimension of X is zero and the degree of X equals δ.*

PROOF. Let A be the homogeneous coordinate ring of X and let ℓ be a linear form which does not vanish at any point of X. If $F \in A_t$ then $\ell F = 0$ in A_{t+1}

if and only if F vanishes at all points of X. It follows that the multiplication by ℓ map $m_\ell : A_t \to A_{t+1}$ is injective so the Hilbert function is non-decreasing. Let $B := A/(\ell)$ and note that since A is a ring generated in degree one, B is also a ring generated in degree one. It follows that $B_s = 0$ implies $B_j = 0$ for all $j \geq s$. By definition of B and the injectivity of m_ℓ we also know that the dimensions of A_j and A_{j+1} coincide if and only if $B_{j+1} = 0$. We conclude that the Hilbert function of A must be strictly increasing forever or become constant after the first point when it is not strictly increasing. For any $t \geq 0$ there is an injective map $A_t \to \mathrm{Fun}(X, \mathbb{C})$ which sends a form $F(x_0, \ldots, x_n)$ to the complex-valued function $\frac{F(x_0,\ldots,x_n)}{\ell(x_0,\ldots,x_n)^t}$ on X. Since every complex-valued function on X can be realized by a polynomial (see Exercise 2.35), this map is surjective for all sufficiently large t. We conclude that the dimension of A_t increases at every stage until it reaches the dimension δ of the space of functions from X to $\mathbb{C}$ at which point it becomes constant. □

EXERCISE 2.35. Suppose $X \subseteq \mathbb{C}^n$ is a finite set and let $g : X \to \mathbb{C}$ be an arbitrary function. Show that there exists a polynomial $p(x_1, \ldots, x_n)$ which agrees with g at all points of X.

DEFINITION 2.36. If X is a finite set then let $\mathrm{reg}(X) := t + 1$ where t is the smallest integer for which $HF_X(t) = \delta$. This is the *Castelnuovo-Mumford regularity of* X, a crucial measure of the "geometric complexity" of a variety.

EXERCISE 2.37. Let X be the set of points $[e_0], [e_1], \ldots, [e_n] \in \mathbb{P}^n$. Prove that X has Castelnuovo-Mumford regularity two. Conversely prove that any set of points which spans $\mathbb{P}^n$ and has regularity two is projectively equivalent to X.

In the remainder of this section we will prove that dimension and degree can both be determined from the geometry of the generic *hyperplane sections* of X (i.e. the intersections of X with generic hyperplanes H), allowing us to relate their geometric and algebraic definitions. The main idea for the proof is the Theorem, due to Bertini, that if H is a generic hyperplane defined by a linear form h then the ideals $I(X \cap H)$ and $I(X) + (h)$ agree for all high enough degrees (see [**17**] for a more precise statement of the Theorem).

THEOREM 2.38. *Suppose X is a variety of dimension d and degree δ. If $d = 0$ then X consists of δ (possibly complex) distinct points. If $d > 0$ and H is a generic hyperplane then $Y := X \cap H$ is a variety of dimension $d - 1$ and degree δ.*

PROOF. If $Y \subseteq X$ then $\mathrm{HF}_Y(t) \leq \mathrm{HF}_X(t)$ for all t and therefore $p_Y(t) \leq p_X(t)$ for all sufficiently large t. By Lemma 2.34 we know that the Hilbert polynomial of a set of k distinct points is the constant k and therefore the Hilbert polynomial of an infinite set X cannot be constant. It follows that every variety of dimension zero is a finite set and our first claim follows from Lemma 2.34.

Since X is a projective variety, no nonzero element of $\mathbb{R}[X]$ is annihilated by the ideal $(X_0, \ldots, X_n)$. It follows that the set of forms in $\mathbb{R}[X]_1$ which are not zero divisors in $\mathbb{R}[X]$ is a nonempty Zariski open set of $\mathbb{P}(\mathbb{R}[X]_1)$. If $h \in \mathbb{R}[X]_1$ is a nonzero divisor then the sequence of vector spaces

$$0 \to \mathbb{R}[X][-1] \to \mathbb{R}[X] \to \mathbb{R}[X]/(h) \to 0$$

is exact in every degree, where $\mathbb{R}[X][-1]$ denotes the module $\mathbb{R}[X]$ graded by $(\mathbb{R}[X][-1])_t = \mathbb{R}[X]_{t-1}$. We conclude that the equality

$$\mathrm{HF}_X(t) - \mathrm{HF}_X(t-1) = \mathrm{HF}_{\mathbb{R}[X]/(h)}(t)$$

holds. By Bertini's Theorem we know that the ideals $I(X \cap V(h))$ and $I(X) + (h)$ coincide in all sufficiently high degrees and therefore the following equality of Hilbert polynomials holds:

$$p_X(t) - p_X(t-1) = p_{X'}(t),$$

where X' is the variety $X \cap V(h)$. If $p_X(t) = a_d t^d + o(t^{d-1})$ for some $d \geq 1$ then we conclude that

$$p_{X'}(t) = d a_d t^{d-1} + o(t^{d-2})$$

holds and therefore X' is a variety of dimension $d-1$ and degree $(d-1)! d a_d = \delta$ as claimed. □

REMARK 2.39. The previous Lemma proves that dimension and degree can be given the geometric definitions from the first paragraph of Section 2.4.1.

More precisely, one can prove that every transverse hyperplane section of an irreducible d-dimensional variety $X \subseteq \mathbb{P}^n$ is a witness for degree, that is:

THEOREM 2.40 (Theorem 5.1 and ensuing comment in [**21**]). *If $X \subseteq \mathbb{P}^n$ is irreducible and d-dimensional and $L \subseteq \mathbb{P}^n$ is a linear subspace of dimension $n-d$ such that:*

(1) *$L \cap X$ is a finite set $\{x_1, \dots, x_k\}$,*
(2) *For every i, the point x_i is a smooth point of X and the tangent spaces $T_{x_i}X$ and $T_{x_i}L$ intersect only at zero*

then $k = \deg(X)$. Moreover a generic $(n-d)$-dimensional linear subspaces satisfies the above properties.

REMARK 2.41. A Theorem of Harris [**13**] implies that moreover the points in the intersection of X and a generic complementary subspace $\{x_1, \dots, x_k\} \subseteq L \cong \mathbb{P}^{n-d}$ are in *uniform position* when X is irreducible. This implies that any subset of $s \leq n-d+1$ such points span a $(s-1)$-dimensional projective space.

2.4.4. *Projections.* In this Section we introduce a collection of maps called projections. We will show that the behavior of our key invariants (and later also of sums of squares and nonnegative polynomials) can often be controlled under projection allowing us to reduce certain geometric questions to the (often simpler) case of hypersurfaces. As a first application of this philosophy we will show that the degree of every irreducible variety not contained in any hyperplane has a universal lower bound. This lower bound leads to the concept of varieties of minimal degree.

If $q \in \mathbb{P}^n$ is a point and $H \subseteq \mathbb{P}^n$ is a hyperplane not containing q then the *projection π_q away from q* is the map $\pi_q : \mathbb{P}^n \setminus \{0\} \to H$ which sends a point $r \neq q$ to the unique point of intersection of the line $\langle r, q\rangle$ which joins r and q and the hyperplane H. In coordinates, if $q = [0 : 0 : \cdots : 0 : 1]$ and $H \cong \mathbb{P}^{n-1}$ is given by $H = V(X_n)$ then $\pi_q([\alpha_0 : \cdots : \alpha_n]) = [\alpha_0 : \cdots : \alpha_{n-1} : 0]$. Often we identify H with $\mathbb{P}^{n-1}$ implicitly and the last component of π is dropped. Note that, up to automorphisms of the image $\mathbb{P}^{n-1}$, the projection is completely determined by the point q (its components are given by any basis for the linear forms vanishing at q and any two such bases differ by a linear change of coordinates).

EXERCISE 2.42. Prove that if $q = [0 : 0 : \cdots : 0 : 1]$ and $H = V(X_n)$ then π_q is given by the formula from the previous paragraph.

If $X \subseteq \mathbb{P}^n$ is a variety then we define its image Y under projection away from q as the Zariski closure $Y := \overline{\pi_q(X \setminus \{q\})}$. For ease of notation we write this as $Y = \pi_q(X)$. As the following exercise shows, the invariants of the projection Y can depend on the relative locations of q and X.

EXERCISE 2.43. Let X be the twisted cubic in $\mathbb{P}^3$. (i.e. the image of the morphism $\nu_2([s:t]) = [s^3 : s^2t : st^2 : t^3]$) and let $Y := \pi_q(X)$. Compute the ideal of Y and prove that:

(1) If $q = [1:0:0:0]$ then $q \in X$ and $\deg(Y) = \deg(X) - 1 = 2$ (inner projection).
(2) If $q = [0:0:1:0]$ then $q \notin X$ and $\deg(Y) = \deg(X) = 3$ (outer projection).

In general, the computation of the ideal of the image of X under π_q can often be reduced to an elimination problem, allowing us, for instance, to compute projections using Gröbner bases. More precisely,

LEMMA 2.44. *If $q = [1:0:\cdots:0]$ and $X \subseteq \mathbb{P}^n$ is the variety defined by the ideal I then the following statements hold:*

(1) *The ideal $J := I \cap \mathbb{C}[X_1, \ldots, X_n]$ defines the variety $Y := \pi_q(X)$ in $\mathbb{P}^{n-1}$*
(2) *The pullback map*
$$\pi^* : \mathbb{C}[Y] = \mathbb{C}[X_1, \ldots, X_n]/J \to \mathbb{C}[X_0, \ldots, X_n]/I = \mathbb{C}[X]$$
is an injective, graded, ring homomorphism.

PROOF. Since X is irreducible $G(X_1, \ldots, X_n) \in I(Y)$ if and only if it vanishes at all points of $\pi_q(X \setminus q)$ if and only if $G \circ \pi_q$ vanishes at all points of X proving that $J = I(Y)$. For the second part note that the map $\mathbb{C}[X_1, \ldots, X_n] \to \mathbb{C}[X_0, \ldots, X_n]/I$ is a graded ring homomorphism and that its kernel consists of polynomials in $X_1, \ldots, X_n$ which vanish on X and therefore equals J proving injectivity as claimed. □

The following Theorem summarizes behavior of the basic invariants of X under inner projections (i.e. projections of X away from a point contained in X). It shows that the behavior of part (1) of the previous exercise is typical for inner projections. Its proof is not self-contained but it contains references to sources where the necessary Lemmas can be found. Recall that the codimension of a variety $X \subseteq \mathbb{P}^n$ is defined as $\operatorname{codim}(X) := n - \dim(X)$.

THEOREM 2.45. *Suppose $X \subset \mathbb{P}^n$ is irreducible and not a linear subspace. If q is a generic point of X and $Y := \pi_q(X)$ then $\dim(Y) = \dim(X)$. If $\operatorname{codim}(X) \geq 2$ then moreover $\deg(Y) = \deg(X) - 1$.*

PROOF. Let L be a generic $(n-d)$-dimensional subspace with $X \cap L = \{x_1, \ldots, x_{\deg(X)}\}$ then let $q = x_1$ and note that $\pi_q : X \to Y$ is generically finite (because the line $\langle q, x_2 \rangle$ intersects X at finitely many points) so $\dim(X) = \dim(Y)$. If X has codimension at least two then $\pi_q : X \to Y$ has degree one (because the line $\langle q, x_2 \rangle$ does not intersect X at any other point of X). By [**21**, Corollary 5.15] the multiplicity of X at q is one and therefore by [**21**, Theorem 5.11] we know that $\deg(X) - 1 = \deg(Y)\deg(\pi_q) = \deg(Y)$ as claimed. □

Applying projections iteratively we obtain the following universal lower bound for the degree of irreducible varieties.

THEOREM 2.46. *If $X \subseteq \mathbb{P}^n$ is an irreducible variety not contained in any hyperplane then* $\deg(X) \geq \operatorname{codim}(X) + 1$.

PROOF. We use induction in the codimension c of X in $\mathbb{P}^n$. If $c = 0$ then the statement holds trivially. If $c = 1$ then X must be a hypersurface of degree ≥ 2 proving the claim. If $c \geq 2$ then let q be a generic point of X and let $Y := \pi_q(X) \subseteq \mathbb{P}^{n-1}$. By Theorem 2.45 X and Y have the same dimension and $\deg(X) - 1 = \deg(Y)$. It follows that $\operatorname{codim}(Y) = \operatorname{codim}(X) - 1$ and the induction hypothesis implies that $\deg(Y) \geq \operatorname{codim}(Y)+1$. We conclude that $\deg(X) \geq \operatorname{codim}(Y)+1+1 = \operatorname{codim}(X) + 1$ as claimed. □

The previous Theorem motivates the following definition, which plays a fundamental role in this Chapter. We will discuss the classification of such varieties in Part 2. A variety $X \subseteq \mathbb{P}^n$ is called non-degenerate if X is not contained in any hyperplane in $\mathbb{P}^n$.

DEFINITION 2.47. A variety $X \subseteq \mathbb{P}^n$ is called of minimal degree if it is irreducible and non-degenerate and $\deg(X) = \operatorname{codim}(X) + 1$.

2.5. On the equality between nonnegative quadrics and sums of squares. As an application of the theory developed so far, we will use Theorem 1.4 to prove Hilbert's characterization of equality between nonnegative forms and sums of squares. The remaining exercises in the Section will ask you to similarly prove that the other characterizations appearing in the introduction are special cases of Theorem 1.4.

PROOF OF THEOREM 1.1 FROM THEOREM 1.4. Let n, d be positive integers and let $X := \nu_d(\mathbb{P}^n) \subseteq \mathbb{P}^k$ where $k = \binom{n+d}{d} - 1$ (see Section 2.3.2 for more information on the Veronese map). Since $\mathbb{P}^n$ is totally real the same is true about X and therefore it follows from Theorem 1.4 that nonnegative quadratic forms on X agree with sums of squares of linear forms on X if and only if X is a variety of minimal degree. Since ν_d is an isomorphism we know that $\dim(X) = n$ and from the exercises we know that $\deg(X) = d^n$. It follows that X is of minimal degree if and only if the following equality holds

$$d^n = \binom{d+n}{n} - n.$$

This occurs if and only if $n = 1$ or $d = 1$ or $(d = 2, n = 2)$ which correspond to the bivariate case, the case of quadratic forms and the case of ternary quartics respectively. □

EXERCISE 2.48. Show that the only solutions to the equation $d^n = \binom{d+n}{n} - n$ are $n = 1$, $d = 1$ or $(d = 2, n = 2)$, finishing the proof of Hilbert's Theorem.

EXERCISE 2.49. *(Segre-Veronese embeddings and bi-forms)* Fix positive integers d_1, d_2, n_1, n_2 and let $\sigma_{d_1,d_2} : \mathbb{P}^{n_1} \times \mathbb{P}^{n_2} \to \mathbb{P}^k$ be the map sending a pair $([\alpha], [\beta])$ to the vector of all products $m(\alpha)g(\beta)$ where m and g are monomials of degrees d_1 and d_2 in the variables $X_0, \ldots, X_{n_1}$ and $Y_0, \ldots, Y_{n_2}$ respectively.

(1) Prove that $X := \sigma_{d_1,d_2}(\mathbb{P}^{n_1} \times \mathbb{P}^{n_2})$ is a variety of minimal degree if and only $d_1 = n_2 = 1$ or $d_2 = n_1 = 1$.
(2) Conclude that the main theorem implies

THEOREM (Choi-Lam-Reznick 1980). *A bi-homogeneous nonnegative polynomial of degree* $(2d_1, 2d_2)$ *in two sets of variables of sizes* $n_1, n_2 \geq 2$ *is a sum of squares of bi-homogeneous forms if and only if either* $(n_1 = 2, d_2 = 1)$ *or* $(d_1 = 1, n_2 = 2)$.

EXERCISE 2.50. *(The homogeneous S-Lemma)*

(1) Prove that the main theorem implies:

THEOREM. *Let* $Q(X_0, \ldots, X_n)$ *be an indefinite quadratic form and let* $Z = V(Q)$. *Every quadratic polynomial* F *which is nonnegative on* Z *is of the form*

$$F = \ell_1^2 + \cdots + \ell_k^2 + \lambda Q$$

for some linear forms $\ell_i(X_0, \ldots, X_n)$ *and some real number* λ.

(2) *Jakubovič's S-Lemma, homogeneous version.* Suppose $H \in \mathbb{R}[X_1, \ldots, X_n]_2$ is a quadratic form and that $\alpha \in \mathbb{R}^n$ satisfies $H(\alpha) > 0$. Use part (1) to prove that if $F \in \mathbb{R}[X_1, \ldots, X_n]_2$ satisfies $F(\beta) \geq 0$ at all points $\beta \in \mathbb{R}^n$ with $H(\beta) \geq 0$ then there exists $\lambda \geq 0$ and linear forms $\ell_1, \ldots, \ell_t \in \mathbb{R}[X_1, \ldots, X_n]_1$ such that

$$F = \ell_1^2 + \cdots + \ell_t^2 + \lambda H$$

(Hint: Consider the form $T := H - X_0^2 \in \mathbb{R}[X_0, \ldots, X_n]_2$ and let $X = V(T)$. Prove that F is a sum of squares on X).

2.6. Appendix to Lecture 1: Optimization and projective geometry. In this Section we show that any polynomial optimization problem over an algebraic set in $\mathbb{R}^n$ is equivalent to one about the cone of nonnegative quadratic forms on some real projective variety. It follows that the projective point of view can be applied to studying any polynomial optimization problem.

More precisely, the most common polynomial optimization problem asks for finding the minimum value of a polynomial function $f(x) \in \mathbb{R}[x_1, \ldots, x_n]$ over a set of the form $Z(\mathbb{R}) := \{\alpha \in \mathbb{R}^n : g_1(\alpha) = \cdots = g_k(\alpha) = 0\}$ for some $g_1, \ldots, g_k \in \mathbb{R}[x_1, \ldots, x_n]$. The optimal value of this optimization problem is given by

(1) $$\lambda^* := \sup \{\lambda \in \mathbb{R} : f(x) - \lambda \text{ is a nonnegative function on } Z(\mathbb{R})\}.$$

In this section we reformulate this problem as one about the cone of nonnegative quadratic forms on a projective variety. To do this we must first clarify the relationship between $\mathbb{C}^n$ and $\mathbb{P}^n$. Let $i : \mathbb{C}^n \to \mathbb{P}^n$ be the map sending a point $(\alpha_1, \ldots, \alpha_n)$ to $[1 : \alpha_1 : \cdots : \alpha_n]$. The map i is injective and its image consists of the open set $X_0 \neq 0$ in $\mathbb{P}^n$. It follows that $\mathbb{P}^n = \mathbb{C}^n \cup \mathbb{P}^{n-1}$ where $\mathbb{P}^{n-1}$ is the hyperplane $V(X_0)$ (called hyperplane at infinity). In words, we have shown that projective space $\mathbb{P}^n$ consists of the points of $\mathbb{C}^n$ together with an additional $\mathbb{P}^{n-1}$ of points "at infinity", one for each direction in $\mathbb{C}^n$.

Having identified the points of $\mathbb{C}^n$ in $\mathbb{P}^n$ we now extend polynomials in $\mathbb{C}^n$ to forms in $\mathbb{P}^n$ via *homogenization*: If $h \in \mathbb{R}[x_1, \ldots, x_n]$ is a (not-necessarily homogeneous) polynomial of degree d then h can be written uniquely as $h = h_d + h_{d-1} + \cdots + h_0$ where the h_j are homogeneous polynomials of degree j. Define the homogenization of h with respect to the variable X_0 as the element of $\mathbb{C}[\mathbb{P}^n]$ given by

$$H := h_d(X_1, \ldots, X_n) + X_0 h_{d-1}(X_1, \ldots, X_n) + \cdots + X_0^d h_0(X_1, \ldots, X_n).$$

If $t = 2s$ is the smallest even degree strictly larger than d define the *even homogenization* of h with respect to X_0, H^{even} as:

$$X_0^{t-d}h_d(X_1,\dots,X_n) + X_0^{t-d-1}h_{d-1}(X_1,\dots,X_n) + \cdots + X_0^t h_0(X_1,\dots,X_n).$$

Note that $H(i(x)) = H^{even}(i(x)) = h(x)$ for all $x \in \mathbb{C}^n$ and that H^{even} has even degree $2s$ and is identically zero on the hyperplane at infinity $V(X_0)$.

THEOREM 2.51. *Let $G_1,\dots,G_k$ be the homogenizations of $g_1,\dots,g_k$ and let $W := V(G_1,\dots,G_K) \subseteq \mathbb{P}^n$. Let F^{even} be the even homogenization of f of degree $t = 2s$ and let $X := \nu_s(W)$. The number λ^* in* (1) *satisfies $\lambda^* := \sup\{\lambda \in \mathbb{R} : A - \lambda Y_I^2 \in P_X\}$ where A is a quadratic form in $\mathbb{P}^k$ with $A \circ \nu_s = F^{even}$ and Y_I is a coordinate in $\mathbb{P}^k$ with $Y_I \circ \nu_s = X_0^s$.*

PROOF. Let λ be any real number. Since the Veronese embedding is an isomorphism defined over $\mathbb{R}$ every real point of X is the image of a real point of $W(\mathbb{R})$. It follows that the quadratic form $A - \lambda Y_I^2$ is nonnegative on $X(\mathbb{R})$ if and only if the degree $2s$ form $L = F^{even} - \lambda X_0^{2s}$ is nonnegative on $W(\mathbb{R})$. The real points $[\alpha] \in W(\mathbb{R}) \subseteq \mathbb{P}^n$ either have $X_0 = 0$ or $X_0 \neq 0$. In the first case $L(\alpha) = 0$ since F^{even} is divisible by X_0. In the second case $[\alpha] = [1 : \alpha_1 : \cdots : \alpha_n]$ for $(\alpha_1,\dots,\alpha_n) \in Z(\mathbb{R}) \subseteq \mathbb{R}^n$ and therefore $L(\alpha)$ has the same sign as $f(\alpha_1,\dots,\alpha_n) - \lambda$ proving the claim. □

REMARK 2.52. The problem of determining $\lambda^* := \sup\{\lambda \in \mathbb{R} : A - \lambda Y_I^2 \in P_X\}$ is a conic optimization problem since it is equivalent to maximizing the linear map $\pi_2 : \mathbb{R}[X]_2 \times \mathbb{R} \to \mathbb{R}$ over the intersection of the convex cone $C = \{(Q,\lambda) : Q \in P_X, \lambda \in \mathbb{R}\}$ and the affine subspace $\{(Q,\lambda) : Q = A - \lambda Y_I^2\}$. In particular it is a semidefinite program whenever P_X is semidefinitely representable (this occurs, for instance when $P_X = \Sigma_X$).

REMARK 2.53. By choosing good equations $g_1,\dots,g_k$ for $Z(\mathbb{R})$ one can choose X to be totally real. This happens for instance if $(g_1,\dots,g_k) = I(Z(\mathbb{R}))$ and W is chosen to be the projective closure of $i(Z(\mathbb{R}))$. There are algorithms for constructing such $g's$ (see [**20**] or [**24**]) and for computing projective closures (see [**10**, Section 8.4]) so this reduction is effective.

3. Main Theorem and the classification of varieties of minimal degree.

Directly contradicting G.C. Rota's good advice [**23**] this Section will have not one, but two main points. The first is the proof of Theorem 1.4 and the second is the statement (although sadly, not its proof) of the classification of varieties of minimal degree, one of the great achievements of the Italian school of algebraic geometry of the late 1800's.

Our proof of Theorem 1.4 will consist of two parts. In Section 3.1 we will show that if $X \subset \mathbb{P}^n$ is irreducible then $\Sigma_X = P_X$ implies that X must be of minimal degree. The key point of the proof of this implication is the fact that equality is preserved under *projections away from real points of* X allowing us to reduce to the hypersurface case. In Section 3.3 we will prove the opposite direction, namely that if $\Sigma_X \neq P_X$ and $X' := X \cap H$ is a general *hyperplane section* of X then $\Sigma_{X'} \neq P_{X'}$. Taking successive hyperplane sections we will be able to reduce the question to the case of points. The proofs presented in these notes have not appeared in print before. The use of projections for the first implication is an idea due to Greg Blekherman which greatly simplifies the arguments from the original article.

Convention. Throughout this Chapter we let $X \subset \mathbb{P}^n$ be a totally real, irreducible and non-degenerate projective variety.

3.1. Main Theorem Part 1: If $P_X = \Sigma_X$ then X must be of minimal degree.

For $p \in X(\mathbb{R})$ let π be the projection away from p and let $Y \subseteq \mathbb{P}^{n-1}$ be the variety given by $Y := \pi_p(X)$. Assume that $P_X = \Sigma_X$. Our first lemma proves that this equality is preserved under such projections.

LEMMA 3.1. *If $\Sigma_X = P_X$ then $\Sigma_Y = P_Y$.*

PROOF. Let $\pi^* : \mathbb{R}[Y] \to \mathbb{R}[X]$ be the pullback map from Lemma 2.44 which is dual to the projection. If $Q \in P_Y$ then $\pi^*(Q)$ is nonnegative in X and therefore $\pi^*(Q) = \sum_{i=1}^t \ell_i^2$ for some linear forms ℓ_i Since $\pi^*(Q)$ vanishes at p we conclude that all the ℓ_i vanish at p and therefore there exist linear forms $\eta_i \in \mathbb{R}[Y]$ such that $\ell_i = \pi^*\eta_i$ for $i = 1, \ldots, t$. Since π^* is an injective ring homomorphism we conclude that $Q = \sum \eta_j^2$ so $Q \in \Sigma_Y$ as claimed. □

The codimension of Y is one less than the codimension of X so applying successive projections we arrive to the case of a hypersurface. Our next Lemma shows that equality fails for most hypersurfaces. Note that the only non-degenerate hypersurfaces of minimal degree are those of degree two.

LEMMA 3.2. *If $X \subseteq \mathbb{P}^n$ is a real hypersurface of degree $d \geq 3$ then $P_X \neq \Sigma_X$.*

PROOF. Since X has degree at least three the ideal $I(X)$ does not contain any quadratic generators and two quadratic forms in $\mathbb{P}^n$ agree in $\mathbb{R}[X]$ if and only if they agree in $\mathbb{R}[\mathbb{P}^n]$. In particular $\Sigma_X = \Sigma_{\mathbb{P}^n}$ and to prove that $P_X \neq \Sigma_X$ it suffices to exhibit a quadratic form which is nonnegative on $X(\mathbb{R})$ and is negative at some point of $\mathbb{P}^n$. Let $i : \mathbb{R}^n \to \mathbb{P}^n$ be the inclusion $i(\alpha_1, \ldots, \alpha_n) = [1 : \alpha_1 : \cdots : \alpha_n]$ and let $W := i^{-1}(X(\mathbb{R}))$. Let $\gamma \in \mathbb{R}^n \setminus W$ and let $\delta > 0$ be a sufficiently small positive real number such that the closed euclidean ball in $\mathbb{R}^n$ centered at γ of radius δ is disjoint from W. Define the quadratic polynomial

$$q = (x_1 - \gamma_1)^2 + \cdots + (x_n - \gamma_n)^2 - \delta$$

and let Q be its homogenization with respect to X_0, that is

$$Q = (X_1 - \gamma_1 X_0)^2 + \cdots + (X_n - \gamma_n X_0)^2 - \delta X_0^2.$$

Note that Q is homogeneous of degree two, is nonnegative on $X(\mathbb{R})$ and has negative sign at $[1 : \gamma_1 : \cdots : \gamma_n]$. We conclude that $Q \in P_X \setminus \Sigma_X$ proving the claim. □

PROPOSITION 3.3. *If X is irreducible and totally real and $\Sigma_X = P_X$ then $X \subset \mathbb{P}^n$ is a variety of minimal degree.*

PROOF. We prove the claim by induction on $c := \operatorname{codim}(X, \mathbb{P}^n)$. The case $c = 1$ is proven in Lemma 3.2. If $X \subseteq \mathbb{P}^n$ has codimension $c > 1$ then by density of $X(\mathbb{R})$ there exists a point $p \in X(\mathbb{R})$ such that the following statements hold:

(1) The subvariety $Y \subseteq \mathbb{P}^{n-1}$ obtained by projecting X away from p is irreducible and non-degenerate, has the same dimension as X and satisfies $\deg(Y) = \deg(X) - 1$.
(2) The set $Y(\mathbb{R})$ is Zariski dense in Y.

Moreover, Lemma 3.1 implies that the equality $\Sigma_Y = P_Y$ holds. Since $\operatorname{codim}(Y, \mathbb{P}^{n-1}) = n - 1 - \dim(Y) = c - 1 < c$ we conclude from the induction hypothesis that Y is a variety of minimal degree. It follows that the equality

$$\deg(X) = \deg(Y) + 1 = \operatorname{codim}(Y, \mathbb{P}^{n-1}) + 1 + 1 = c - 1 + 1 + 1 = c + 1$$

holds, proving the proposition. □

3.2. The many faces of the cone Σ_X^* dual to Σ_X. Next we would like to prove that $\Sigma_X = P_X$ for varieties X of minimal degree. In order to do this we will study general properties of the cone $\Sigma_X^* \subseteq \mathbb{R}[X]_2^*$ which is the convex cone dual to $\Sigma_X \subseteq \mathbb{R}[X]_2$.

We begin by recalling the definition of dual convex cone. If V is a real finite-dimensional vector space and $C \subseteq V$ is a convex cone then the *dual cone* to C is the convex cone in the dual vector space V^* given by

$$C^* := \{\phi \in V^* : \forall c \in C\, (\phi(c) \geq 0)\}.$$

The dual cone is extremely useful because its points give a description of the closure $\overline{C}$ (in the usual euclidean topology of V) via linear inequalities. The fundamental bi-duality Theorem of convex geometry [**2**] says that if $z \in V$ is any point not in $\overline{C}$ then this fact can always be witnessed by some element C^* in the sense that there exists $\phi \in C^*$ such that $\phi(z) < 0$ and $\phi(C) \geq 0$. From this point of view it is not surprising that a good way (strictly speaking the *only way*) to distinguish sums of squares from non sums of squares is by understanding the cone Σ_X^*.

We begin by proving the easy (but fundamental) fact that Σ_X^* is a spectrahedron and then the deeper fact that it has *many* distinct facets in any euclidean neighborhood of most points of its boundary whenever $X(\mathbb{R})$ is dense in X. For such varieties, an affine slice of the cone Σ_X^* is therefore much more like a euclidean ball than like a polytope.

Definition 3.4. The cone $\Sigma_X^* \subseteq \mathbb{R}[X]_2^*$ dual to the sums of squares on X is given by

$$\Sigma_X^* := \left\{\ell \in \mathbb{R}[X]_2^* : \forall q \in \mathbb{R}[X]_1 \left(\ell(q^2) \geq 0\right)\right\}.$$

Recall that the quotient map defines a surjection $r : \mathbb{R}[\mathbb{P}^n]_2 \to \mathbb{R}[X]_2$. Its transpose $r^* : \mathbb{R}[X]_2^* \to \mathbb{R}[\mathbb{P}^n]_2^*$ is therefore an injective linear map and thus defines an isomorphism of cones between Σ_X^* and its image $r^*(\Sigma_X^*)$. The map r^* allows us to identify Σ_X^* with a convex set of symmetric matrices and to think in a very concrete manner about this object. In particular, if $\ell \in \Sigma_X^*$ then we define its rank (resp. its co-rank) as the rank (resp $(n+1)$ minus the rank of) the symmetric matrix $r^*(\ell)$.

Example 3.5. If $X \subseteq \mathbb{P}^3$ is the twisted cubic curve then Σ_X^* is the set of 4×4 symmetric PSD matrices which annihilate the ideal

$$I(X) := \left(2 \times 2 \text{ minors of } \begin{pmatrix} X_0 & X_1 & X_2 \\ X_1 & X_2 & X_3 \end{pmatrix} \right).$$

For instance annihilating $X_0X_2 - X_1^2 \in I(X)$ means that the coefficients must satisfy $A_{02} - A_{11} = 0$. Simplifying these equations we obtain that for the twisted cubic

$$\Sigma_X^* := \left\{ \begin{pmatrix} A & B & C & D \\ B & C & D & E \\ C & D & E & F \\ D & E & F & G \end{pmatrix} \text{ which are PSD and have real entries.} \right\}.$$

EXERCISE 3.6. Prove that the cone Σ_X^* is a spectrahedral cone. More precisely, show that $r^*(\Sigma_X^*)$ is equal to the intersection of a linear space and the cone of positive semidefinite quadratic forms in $\mathbb{R}[\mathbb{P}^n]_2^*$.

EXERCISE 3.7. Describe the spectrahedral cone Σ_X^* when:

(1) $X \subseteq \mathbb{P}^n$ is a rational normal curve.
(2) $X \subseteq \mathbb{P}^5$ is the Veronese surface.

The following Theorem summarizes several key characteristics of the convex geometry of the cones Σ_X^* whenever $X(\mathbb{R})$ is Zariski-dense in X. Note that, unlike all previous statements in these notes, those in the following Theorem refer to the usual metric topology in the real vector space $\mathbb{R}[X]_2$ and not to the Zariski topology,

THEOREM 3.8. [**7**, Corollary 2.8] *If $X(\mathbb{R})$ is Zariski-dense in X then Σ_X is closed and Σ_X^* is* **many-faceted**, *meaning that it satisfies the following two properties (in the euclidean topology of $\mathbb{R}[X]_2^*$):*

(1) *The points of co-rank one are dense in the topological boundary $\partial\Sigma_X^*$ of Σ_X^*.*
(2) *If $\ell \in \partial\Sigma_X^*$ is a point of co-rank one then there exists a euclidean ball B centered at ℓ such that $B \cap \partial\Sigma_X^*$ consists exclusively of points of co-rank one. Moreover the image of the map $T : B \cap \partial\Sigma_X^* \to \mathbb{P}^n(\mathbb{R})$ sending $\hat{\ell}$ to its kernel is open at ℓ (i.e. any neighborhood $U \ni \ell$ is mapped to a set $T(U)$ which contains some euclidean open set around $T(\ell)$ in $\mathbb{P}^n(\mathbb{R})$).*

3.3. Main Theorem Part 2: If X is of minimal degree then $P_X = \Sigma_X$. In this section we will prove that $P_X = \Sigma_X$ for varieties X of minimal degree. The key point of the proof is showing that, whenever $P_X \neq \Sigma_X$ the same is true for sufficiently general hyperplane sections $Y := X \cap H$ of X. The main tool for establishing this implication is Theorem 3.8. Applying this reduction $\dim(X)$-times we reach the case of points where our assumption leads to a contradiction as the following Lemma shows,

LEMMA 3.9. *If $X \subseteq \mathbb{P}^m$ is a non-degenerate set of $m+1$ points then $P_X = \Sigma_X$.*

PROOF. Since X is non-degenerate, any set of affine representatives for X forms a basis of $\mathbb{C}^{m+1}$. In particular we can choose coordinates for which our points are $[e_0], \ldots, [e_m]$. If $H \in P_X$ then the numbers $A_i = H(e_i)$ are nonnegative and therefore the form $Q := \sum \left(\sqrt{A_i} X_i\right)^2$ is a sum of squares in $\mathbb{R}[X]$. Since $Q(e_i) = A_i$ for all i the equality $Q = H$ holds in $\mathbb{R}[X]$ and therefore $Q \in \Sigma_X$ as claimed. □

THEOREM 3.10. *If X is a variety of minimal degree for which $X(\mathbb{R})$ is Zariski-dense then $\Sigma_X = P_X$.*

PROOF. Suppose not and let $g \in P_X \setminus \Sigma_X$ be a nonnegative quadratic form which is not a sum of squares in X. By Theorem 3.8 the cone Σ_X is closed and therefore there exists a strict separator $\ell \in \Sigma_X^*$ with $\ell(\Sigma_X) \geq 0$ and $\ell(g) < 0$.

We can choose such a separator in the boundary of Σ_X^* and by property (1) of Theorem 3.8 we can choose it to be of co-rank one. The locus of hyperplanes H for which $X \cap H$ is a variety of dimension $d-1$ and of degree $\deg(X)$ (irreducible if X has dimension at least two) is Zariski open. We conclude that there exists an $\ell' \in \Sigma_X^*$ and real linear form h such that $\ell'(h^2) = 0$ and such that $Y := X \cap V(h)$ is a variety of dimension $d-1$, of minimal degree in $\mathbb{P}^{n-1}$ whose set of real points is Zariski dense. Since $(h)_2$ is annihilated by ℓ' the function ℓ' induces a linear function $\ell_2 : \mathbb{R}[Y]_2 = (\mathbb{R}[X]/(h))_2 \to \mathbb{R}$ with $\ell_2(\Sigma_Y) \geq 0$ and $\ell_2(g) < 0$ proving that $P_Y \neq \Sigma_Y$. Applying this procedure iteratively $\dim(X)$ many times we end up with a finite, non-degenerate set of points Z of cardinality $c+1$ in $\mathbb{P}^c$ where $c = \operatorname{codim}(X)$ with the property that $\Sigma_Z \neq P_Z$, contraddicting Lemma 3.9. We conclude that $\Sigma_X = P_X$ as claimed. □

3.4. The classification of varieties of minimal degree. So far we have shown that Theorem 1.4 implies several known characterizations of equality between nonnegative polynomials and sums of squares. In order to produce other characterizations we would like to understand all varieties of minimal degree. In this Section we will describe their classification due to Del Pezzo and Bertini [**3**]. For a proof (valid in all characteristics) the reader is referred to Eisenbud and Harris [**12**]. We begin by stating the Theorem and then we will discuss the varieties appearing in the classification. A final application to sums of squares appears at the end of the section.

THEOREM 3.11. *Let $X \subseteq \mathbb{P}^n$ be irreducible and not contained in any hyperplane in $\mathbb{P}^n$. If X is of minimal degree (i.e. $\deg(X) = \operatorname{codim}(X) + 1$) then either:*

(1) *$X = \mathbb{P}^n$ or*
(2) *X is a quadric hypersurface or*
(3) *X is a cone over the Veronese surface in $\mathbb{P}^5$ or*
(4) *X is a rational normal scroll.*

To clarify the statement recall that a point $q \in Y \subseteq \mathbb{P}^n$ is a cone point of a variety Y if for every $y \in Y$ the line $\langle q, y\rangle \subseteq Y$. The cone over a variety Z with apex q outside the span of Z is the union of all lines $\langle q, z\rangle$ for $z \in Z$. By a cone over X we mean the result of applying any finite number of such coning constructions. In coordinates, if $q = [0 : 0 : \cdots : 0 : 1]$ then q is a cone point of Y if $I(Y)$ has a generating set consisting of polynomials which do not involve the last variable X_n. If $X := Y \cap V(X_n)$ then Y is a cone over X with apex q.

EXERCISE 3.12. (Some properties of cones)

(1) Prove that $q = [0 : 0 : \cdots : 0 : 1] \in Y$ is a cone point of Y if and only if $I(Y)$ has a generating set consisting of polynomials which do not involve the variable X_n.
(2) Determine the dimension and degree of a cone $C \subseteq \mathbb{P}^{n+k}$ over a variety $Z \subseteq \mathbb{P}^n$ in terms of those of Z.
(3) Write the ideal of a variety $Y \subseteq \mathbb{P}^6$ which is a cone over the Veronese surface in $\mathbb{P}^5$.

Next we discuss what rational normal scrolls are. These varieties occur throughout projective algebraic geometry (and according to M. Reid "the student will never regret the investment of time studying them"). We will give the most concrete definition we know and then summarize some of their basic properties. The interested reader should consult [**22**, Chapter 2] for surface scrolls and [**14**] for further information.

DEFINITION 3.13. Given integers $0 \leq a_1 \leq a_2 \leq \cdots \leq a_m$ define the Lawrence prism $L(a_1, \ldots, a_m)$ to be the lattice polytope given by

$$\operatorname{Conv}(0, e_1, \ldots, e_{m-1}, a_1 e_m, e_1 + a_2 e_m, \ldots, e_{m-1} + a_m e_m) \subseteq \mathbb{R}^m.$$

The prism $L(a_1, \ldots, a_m)$ is the result of using the standard unit simplex $\operatorname{Conv}(0, e_1, \ldots, e_{m-1})$ as base and putting heights $a_1, \ldots, a_m$ in the e_m direction above its vertices. The lattice points of L are precisely those contained in each of the m vertical lines and therefore we have $N := \sum_{i=1}^m (a_i + 1) = m + \sum_{i=1}^m a_i$ such points. Each such lattice point $(n_1, \ldots, n_m)$ corresponds uniquely to a monomial $x_1^{n_1} x_2^{n_2} \ldots x_m^{n_m}$.

DEFINITION 3.14. The rational normal scroll $S(a_1, \ldots, a_m)$ is the closure of the image of $(\mathbb{C}^*)^m$ under the map $\phi : (\mathbb{C}^*)^m \to \mathbb{P}^{N-1}$ which sends a point β to $[m_1(\beta) : \cdots : m_N(\beta)]$ where the m_i are an enumeration of the monomials corresponding to the lattice points of $L(a_1, \ldots, a_m)$. The scroll $S(a_1, \ldots, a_m) \subseteq \mathbb{P}^{N-1}$ has dimension m and degree $N - m = \sum a_i$. It is known that they are always defined by quadratic equations.

EXAMPLE 3.15. The Lawrence prism $L(1, 2)$ is the convex hull of $(0, 0), (1, 0), (0, 1), (1, 2)$. The scroll $X := S(1, 2)$ is therefore the image of the map

$$[A : B : C : D : E] = [1 : y : x : xy : xy^2]$$

There are 12 lattice points in $2L(1, 2)$ and therefore the ideal of X contains $15 - 12 = 3$ linearly independent quadrics. By inspecting the map we see that these quadrics are $AD - BC, BD - AE, D^2 - EC$ which must define the ideal.

REMARK 3.16. When finding the equations of the image we want a subvariety *of projective space* and thus are interested only in homogeneous equations (or rather in equations whose homogeneous components belong to the ideal). In particular the equation $D - BC$ is not in the ideal of the image.

More geometrically, one can give a construction of the scroll $S(a_1, \ldots, a_m)$ as follows:

(1) Choose subspaces $\mathbb{P}^{a_1}, \ldots, \mathbb{P}^{a_m} \subseteq \mathbb{P}^{N-1}$ which are pairwise disjoint and span $\mathbb{P}^{N-1}$ and for $i = 1, \ldots, m$ let $C_i \subseteq \mathbb{P}^{a_i}$ be a chosen rational normal curve.
(2) Fix isomorphisms $\phi_i : C_1 \to C_i$ for $i = 2, \ldots, m$.
(3) The rational normal scroll $S(a_1, \ldots, a_m)$ is the variety defined as the union of the $(m-1)$-dimensional projective subspaces spanned by m-tuples of points identified by the ϕ_i, that is:

$$S(a_1, \ldots, a_t) = \overline{\bigcup_{\alpha \in C_1} \langle \alpha, \phi_2(\alpha), \ldots, \phi_m(\alpha) \rangle}$$

REMARK 3.17. The previous construction when $m = 2$ motivates the term scroll. To see this, think of the rational normal curves as the two ends of a roll of parchment (the scroll).

Rational normal scrolls admit determinantal equations which we now describe. Suppose $X_0^{(i)}, \dots, X_{a_i}^{(i)}$ are coordinates for $\mathbb{P}^{a_i} \subseteq \mathbb{P}^{N-1}$ in the above decomposition. Then the ideal of $S(a_1, \dots, a_m)$ is given by the 2×2 minors of the matrix

$$\begin{pmatrix} X_0^{(1)} & X_1^{(1)} & \dots & X_{a_1-1}^{(1)} & \dots & X_0^{(m)} & X_1^{(m)} & \dots & X_{a_m-1}^{(m)} \\ X_1^{(1)} & X_2^{(1)} & \dots & X_{a_1}^{(1)} & \dots & X_1^{(m)} & X_2^{(m)} & \dots & X_{a_m}^{(m)} \end{pmatrix}$$

In particular we see that such rational normal scrolls X are real. Since the real points are Zariski-dense in $(\mathbb{C}^*)^m$ we conclude that X is totally real.

EXERCISE 3.18. Prove that:

(1) The scrolls defined from Lawrence prisms satisfy the previous geometric construction.
(2) The 2×2 minors above are contained in the ideal of rational normal scrolls as defined by Lawrence prisms.

EXERCISE 3.19. Describe the spectrahedral cone Σ_X^* when $X = S(a_1, \dots, a_m)$. Prove that every extreme ray of Σ_X^* has rank one.

From our description of scrolls via Lawrence prisms and Theorem 1.4 we obtain the following consequence,

THEOREM 3.20. *Suppose $F(x_1, \dots, x_m)$ is a nonnegative polynomial supported in $2L$ for some Lawrence prism $L = (a_1, \dots, a_m)$ (i.e. a polynomial whose monomials correspond to lattice points in $2L$). The polynomial F is nonnegative if and only if it can be written as a sum of squares of polynomials supported in L.*

EXERCISE 3.21. Prove Theorem 3.20 from the main Theorem.

It is natural to ask whether the hypotheses in Theorem 1.4 can be weakened. A possible extension is to consider *reducible* totally real varieties. Theorem 1.4 can then be generalized by replacing varieties of minimal degree with their reducible counterpart, known as 2-regular varieties. The interested reader should refer to [**5**], where it is shown that:

THEOREM 3.22. [**5**] *Let $X \subset \mathbb{P}^n$ be a real and non-degenerate projective variety such that the set $X(\mathbb{R})$ of real points is Zariski dense. Every nonnegative real quadratic form on X is a sum of squares of linear forms in $\mathbb{R}[X]$ if and only if X is 2-regular.*

Acknowledgments. We wish to thank Greg Blekherman, Rainer Sinn and Greg G. Smith for many useful conversations during the completion of these notes. In particular we thank Greg Blekherman for the key argument in Section 3.1. We also wish to thank Jose Darío Bastidas for comments on a preliminary version of this document.

We wish to thank R. Thomas, P.Parrilo, the AMS and the MAA for the opportunity of being one of the lecturers in the Short course on Sums of Squares: Theory and Applications during the Joint Mathematics Meetings in Baltimore.

We also like to thank R. Thomas, P.Parrilo and the anonymous reviewers for their careful reading of our manuscript and their many insightful suggestions.

M. Velasco is partially supported by research funds from Universidad de los Andes, Facultad de Ciencias, Proyecto INV-2018-50-1392 and by travel funding from the AMS during the Joint Mathematics Meetings in Baltimore.

References

[1] M. F. Atiyah and I. G. Macdonald, *Introduction to commutative algebra*, Student economy edition, Addison-Wesley Series in Mathematics, Westview Press, Boulder, CO, 2016. For the 1969 original see [MR0242802]. MR3525784

[2] Alexander Barvinok, *A course in convexity*, Graduate Studies in Mathematics, vol. 54, American Mathematical Society, Providence, RI, 2002. MR1940576

[3] E. Bertini, *Introduzione alla geometria proiettiva degli iperispazi*, 1907.

[4] Grigoriy Blekherman, Pablo A. Parrilo, and Rekha R. Thomas (eds.), *Semidefinite optimization and convex algebraic geometry*, MOS-SIAM Series on Optimization, vol. 13, Society for Industrial and Applied Mathematics (SIAM), Philadelphia, PA; Mathematical Optimization Society, Philadelphia, PA, 2013. MR3075433

[5] Grigoriy Blekherman, Rainer Sinn, and Mauricio Velasco, *Do sums of squares dream of free resolutions?*, SIAM J. Appl. Algebra Geom. **1** (2017), no. 1, 175–199, DOI 10.1137/16M1084560. MR3633773

[6] Grigoriy Blekherman, Gregory G. Smith, and Mauricio Velasco, *Sums of squares and varieties of minimal degree*, J. Amer. Math. Soc. **29** (2016), no. 3, 893–913, DOI 10.1090/jams/847. MR3486176

[7] Grigoriy Blekherman, Gregory G. Smith, and Mauricio Velasco, *Sharp degree bounds for sum-of-squares certificates on projective curves* (English, with English and French summaries), J. Math. Pures Appl. (9) **129** (2019), 61–86, DOI 10.1016/j.matpur.2018.12.010. MR3998790

[8] Jacek Bochnak, Michel Coste, and Marie-Françoise Roy, *Real algebraic geometry*, Ergebnisse der Mathematik und ihrer Grenzgebiete (3) [Results in Mathematics and Related Areas (3)], vol. 36, Springer-Verlag, Berlin, 1998. Translated from the 1987 French original; Revised by the authors. MR1659509

[9] M. D. Choi, T. Y. Lam, and B. Reznick, *Sums of squares of real polynomials*, K-theory and algebraic geometry: connections with quadratic forms and division algebras (Santa Barbara, CA, 1992), Proc. Sympos. Pure Math., vol. 58, Amer. Math. Soc., Providence, RI, 1995, pp. 103–126. MR1327293

[10] David A. Cox, John Little, and Donal O'Shea, *Ideals, varieties, and algorithms: An introduction to computational algebraic geometry and commutative algebra*, 4th ed., Undergraduate Texts in Mathematics, Springer, Cham, 2015. MR3330490

[11] David Eisenbud, *Commutative algebra: With a view toward algebraic geometry*, Graduate Texts in Mathematics, vol. 150, Springer-Verlag, New York, 1995. MR1322960

[12] David Eisenbud and Joe Harris, *On varieties of minimal degree (a centennial account)*, Algebraic geometry, Bowdoin, 1985 (Brunswick, Maine, 1985), Proc. Sympos. Pure Math., vol. 46, Amer. Math. Soc., Providence, RI, 1987, pp. 3–13, DOI 10.1090/pspum/046.1/927946. MR927946

[13] Joe Harris, *The genus of space curves*, Math. Ann. **249** (1980), no. 3, 191–204, DOI 10.1007/BF01363895. MR579101

[14] Joe Harris, *Algebraic geometry: A first course*, Graduate Texts in Mathematics, vol. 133, Springer-Verlag, New York, 1995. Corrected reprint of the 1992 original. MR1416564

[15] David Hilbert, *Ueber die Darstellung definiter Formen als Summe von Formenquadraten* (German), Math. Ann. **32** (1888), no. 3, 342–350, DOI 10.1007/BF01443605. MR1510517

[16] V. A. Jakubovič, *The S-procedure in nonlinear control theory* (Russian, with English summary), Vestnik Leningrad. Univ. **1** (1971), 62–77. MR0297429

[17] Jean-Pierre Jouanolou, *Théorèmes de Bertini et applications* (French), Progress in Mathematics, vol. 42, Birkhäuser Boston, Inc., Boston, MA, 1983. MR725671

[18] George R. Kempf, *Algebraic varieties*, London Mathematical Society Lecture Note Series, vol. 172, Cambridge University Press, Cambridge, 1993. MR1252397

[19] Jean Bernard Lasserre, *Moments, positive polynomials and their applications*, Imperial College Press Optimization Series, vol. 1, Imperial College Press, London, 2010. MR2589247

[20] Jean-Bernard Lasserre, Monique Laurent, Bernard Mourrain, Philipp Rostalski, and Philippe Trébuchet, *Moment matrices, border bases and real radical computation*, J. Symbolic Comput. **51** (2013), 63–85, DOI 10.1016/j.jsc.2012.03.007. MR3005782

[21] David Mumford, *Algebraic geometry. I: Complex projective varieties*, Vol. 221, Springer-Verlag, Berlin-New York, 1976. MR0453732

[22] Miles Reid, *Chapters on algebraic surfaces*, Complex algebraic geometry (Park City, UT, 1993), IAS/Park City Math. Ser., vol. 3, Amer. Math. Soc., Providence, RI, 1997, pp. 3–159. MR1442522

[23] Gian-Carlo Rota, *Ten lessons I wish I had been taught*, Notices Amer. Math. Soc. **44** (1997), no. 1, 22–25. MR1419919

[24] S. J. Spang, *On the computation of the real radical*, 2007.

Departamento de matemáticas, Universidad de los Andes, Carrera 1[RA]#18A-12, Bogotá, Colombia

Email address: mvelasco@uniandes.edu.co

Proceedings of Symposia in Applied Mathematics
Volume **77**, 2020
https://doi.org/10.1090/psapm/077/00684

Sums of squares in theoretical computer science

Ankur Moitra

ABSTRACT. When faced with a computationally hard optimization problem, the Sum-of-Squares hierarchy yields a sequence of algorithms that trade off complexity with accuracy. However in many cases, understanding exactly how powerful these algorithms are turns out to be quite challenging. In this chapter, we give an introduction to both upper and lower bounds. Through examples, we will explain how to round the higher levels of the Sum-of-Squares hierarchy and how to analyze the resulting algorithms. Finally, we will show how to prove lower bounds against the Sum-of-Squares hierarchy for random constraint satisfaction problems and discuss recent progress on the planted clique problem.

This chapter will give an introduction to how to use the Sum-of-Squares hierarchy to design approximation algorithms, as well as how to prove lower bounds against it. This chapter will be far from comprehensive. For a much more detailed introduction to the theoretical computer science perspectives on Sum-of-Squares see Barak and Steurer's excellent book [**9**].

1. MAXCUT

We begin by showing how the Sum-of-Squares hierarchy can be used to give efficiently computable convex programming relaxations to NP-hard optimization problems. We will start with the famous MAXCUT problem:

DEFINITION 1.1. Given an unweighted graph $G = (V, E)$, the goal is to find a subset $U \subset V$ that maximizes $|E(U, V \setminus U)|$, where $E(S, T)$ is the set of edges with one endpoint in S and the other endpoint in T.

This is a classic NP-hard optimization problem. In fact, it was on Karp's [**16**] original list of 21 NP-complete problems.[1] You can also think of this problem as asking: What is the largest subgraph of G (where we count the number of edges included in the subgraph) that is bipartite?

Since MAXCUT is NP-hard, we will focus on designing approximation algorithms for it. More precisely, our goal is to give an algorithm that finds a cut $U \subset V$ where $|E(U, V \setminus U)|$ is at least an α factor times the optimal value. We

2010 *Mathematics Subject Classification.* Primary 68R01, 68Q01.
This work was supported in part by NSF grant CCF-1565235.

[1] More precisely, he showed that the weighted version was NP-hard and this was relaxed to the unweighted problem in later work.

call this an α-approximation algorithm. First, we note that it is trivial to get an 1/2-approximation algorithm:

CLAIM 1.2. There is a randomized polynomial time 1/2-approximation algorithm for MAXCUT.

PROOF. We can construct U randomly as follows: For each node in V, include it in U with probability 1/2 and otherwise leave it out. Do this independently for each node. Then it is easy to see that the probability that any edge (u, v) ends up crossing the cut—i.e. having one endpoint in U and the other in $V \setminus U$—is exactly 1/2. And now by linearity of expectation, we have that the expected number of edges in $|E(U, V \setminus U)|$ is at least $|E|/2$, and certainly the optimum is at most $|E|$. □

For a long time, this was the best known approximation algorithm. Even various linear programming relaxations turn out to not do any better.

In a seminal work, that introduced semidefinite programming into approximation algorithms, Goemans and Williamson [**12**] gave a α-approximation algorithm for $\alpha \geq 0.878$. While their relaxation and its analysis does not need any of the machinery of Sum-of-Squares, it turns out to be a great place to start—both in understanding how Sum-of-Squares gives a general family of relaxations for various NP-hard graph partitioning problems, and also in how to round a feasible solution in the semidefinite program into a feasible solution for the original problem. This latter part will be surprisingly subtle when we move beyond MAXCUT, and we will elaborate on this later.

We will start by formulating MAXCUT as a polynomial optimization problem. We can write it as

$$\begin{aligned} \max \quad & \sum_{(i,j)\in E} (x_i - x_j)^2 \\ \text{subject to} \quad & x_i^2 = x_i, \text{ for all } i \end{aligned}$$

The constraint $x_i^2 = x_i$ implies that x_i is either zero or one. Then we can think of the set of x_i's that are one as the set U, and the objective function contributes one for each edge in $E(U, V \setminus U)$ and zero otherwise.

Traditionally, the Sum-of-Squares hierarchy is introduced as a relaxation for solving polynomial optimization problems over polynomial constraints. Here we will take a dual view. Instead of thinking of it as a proof system that becomes more powerful as we increase the degree of the proofs that we allow, we will think of it as producing an operator that we will call a *pseudo-expectation* operator. You should think of this as acting like a distribution over solutions—which in our case are assignments to the variables that we can map into cuts.

DEFINITION 1.3. A degree d pseudo-expectation $\widetilde{\mathbb{E}}$ is an operator

$$\widetilde{\mathbb{E}} : \mathcal{P}_n^{\leq d} \to \mathbb{R}$$

where $\mathcal{P}_n^{\leq d}$ is the set of degree at most d polynomials in variables $x_1, x_2, \cdots, x_n$. The operator satisfies the following conditions:

(1) $\widetilde{\mathbb{E}}$ is linear
(2) $\widetilde{\mathbb{E}}[1] = 1$
(3) $\widetilde{\mathbb{E}}[p^2] \geq 0$ for any polynomial p of degree at most $d/2$.

The above constraints are not problem specific. Often we add some additional problem specific constraints. In the case of MAXCUT we will also enforce:

(4) $\widetilde{\mathbb{E}}[x_i^2 p] = \widetilde{\mathbb{E}}[x_i p]$ for any polynomial p of degree at most $d-2$.

When the operator $\widetilde{\mathbb{E}}$ satisfies all of the above constraints, we will say that it is a degree d pseudo-expectation for MAXCUT. Now we can write down our semidefinite programming relaxation for MAXCUT:

$$\begin{aligned} \max \quad & \widetilde{\mathbb{E}}\Big[\sum_{(i,j)\in E} (x_i - x_j)^2\Big] \\ \text{subject to} \quad & \widetilde{\mathbb{E}} \text{ is a degree } d \text{ pseudo-expectation for MAXCUT} \end{aligned}$$

Now, how should you think about this complicated object? What it's trying to do is act like a distribution on feasible solutions. So let's see how a distribution on feasible solutions leads to a pseudo-expectation operator:

LEMMA 1.4. *If there is a distribution on cuts $U \subset V$ where the expected value of $|E(U, V \setminus U)|$ is at least k, then there is a degree d pseudo-expectation that satisfies*

$$\widetilde{\mathbb{E}}\Big[\sum_{(i,j)\in E} (x_i - x_j)^2\Big] \geq k$$

PROOF. We can construct the pseudo-expectation operator explicitly. For any polynomial $p = p(x_1, \cdots, x_n)$ we will set

$$\widetilde{\mathbb{E}}[p] = \mathbb{E}_U[p(1_{x_1 \in U}, \cdots, 1_{x_n \in U})]$$

We leave it as an exercise to the reader to check that all of the conditions in Definition 1.3 are satisfied, and that the objective value is at least k. □

Now if we could maximize the polynomial $\sum_{(i,j)\in E}(x_i - x_j)^2$ over all cuts, we would be done. We could solve MAXCUT optimally and $P = NP$ and we could all go home. But there are some issues. You cannot even write down a distribution on cuts succinctly because you would need to specify 2^n values, one for each cut. What the pseudo-expectation does to circumvent this is it only asks you to write down enough information about the distribution so that we can evaluate it on degree up to d polynomials. More precisely, knowing that the expectation is linear (and same with the pseudo-expectation) you could write down one number for each of the $O(n^d)$ monomials.

The rest of the constraints in the pseudo-expectation are just safety checks. If $\widetilde{\mathbb{E}}$ really were the operator that maps a polynomial to its expectation under a distribution on cuts, then certainly $\widetilde{\mathbb{E}}[1] = 1$. Also $\widetilde{\mathbb{E}}[x_i^2 p] = \widetilde{\mathbb{E}}[x_i p]$ because the distribution only places non-zero probability on assignments to the variables x_i that are zero or one valued. The most powerful constraint is that $\widetilde{\mathbb{E}}[p^2] \geq 0$. This is both what leads to state-of-the-art algorithms but also the hardest condition to verify when you want to prove lower bounds—i.e. when you want to construct a feasible solution whose objective value is much larger than the optimal value of the original polynomial optimization problem.

2. Rounding SOS for MAXCUT

We will show the following theorem:

THEOREM 2.1. *Any feasible solution to the degree 2 Sum-of-Squares relaxation for MAXCUT that achieves objective value k can be rounded to a random cut U*

where

$$\mathbb{E}[|E(U, V \setminus U)|] \geq \alpha_{GW} k$$

where $\alpha_{GW} \triangleq \min_{-1 \leq \rho \leq 1} \frac{2 \arccos \rho}{(1-\rho)\pi} \geq 0.878$.

The constant above is often called the Goemans-Williamson [**12**] constant. It turns out to be at the heart of some of the deepest mysteries in theoretical computer science, like the validity of the *Unique Games Conjecture.*

The main question is: How do we round a pseudo-expectation to find a cut? In the case of a degree 2 pseudo-expectation this turns out to be quite easy. But even for degree 4 the fundamental approach we are using will not work. With this as a caveat, the idea is to sample from a Gaussian distribution whose moments match the pseudo-moments (that is, the evaluation of the pseudo-expectation operator on degree up to two monomials).

CLAIM 2.2. For a feasible solution to the Sum-of-Squares relaxation for MAX-CUT, we can assume that for all i

$$\widetilde{\mathbb{E}}[x_i] = 1/2$$

without loss of generality.

The intuition for this claim is that the cuts U and $V \setminus U$ have the same objective value. So we might as well as put the same probability on both, in which case for each node i, the probability it is in U is exactly $1/2$. Anyways, the useful information about the cut that pertains to its objective value is the correlation between pairs of nodes of which side they belong to.

PROOF. Given a pseudo-expectation operator $\widetilde{\mathbb{E}}$ we can construct a new pseudo-expectation operator $\widetilde{\mathbb{E}}'$ as

$$\widetilde{\mathbb{E}}'[p(x_1, \cdots, x_n)] = \frac{1}{2}\widetilde{\mathbb{E}}[p(x_1, \cdots, x_n)] + \frac{1}{2}\widetilde{\mathbb{E}}[p(1 - x_1, \cdots, 1 - x_n)]$$

Again, we leave it as an exercise to the reader to check that if $\widetilde{\mathbb{E}}$ satisfies the conditions of being a degree 2 pseudo-expectation operator then $\widetilde{\mathbb{E}}'$ does too. Moreover one can also check that $\widetilde{\mathbb{E}}'$ achieves the same objective value. □

Now we get to the most interesting part of the proof. We can sample a Gaussian random variable y in n dimensions with mean μ and covariance Σ where

$$\mu = \widetilde{\mathbb{E}}[x] \text{ and } \Sigma = \widetilde{\mathbb{E}}[(x - \mu)(x - \mu)^T]$$

where x is a vector with variables $x_1, \cdots, x_n$. Here we have applied the pseudo-expectation operator to a matrix and by this we mean applying it entry-wise. Observe that for any vector c we have

$$c^T \Sigma c = \widetilde{\mathbb{E}}[p(x)^2]$$

for some linear $p(x)$. The right hand side is nonnegative by constraint (3) and hence Σ is positive semidefinite.

Finally we construct the set U by including nodes for which $y_i \leq 1/2$ and leaving out the rest. This is called Gaussian rounding. We will show that

$$\mathbb{E}[|E(U, V \setminus U)|] \geq \alpha_{GW} k$$

where the expectation is over the random vector y and k is the objective value of the pseudo-expectation. We will break up the analysis into two parts. First we will calculate the contribution of an edge (i, j) to the objective of the relaxation. For notational convenience, we define

$$\rho_{i,j} \triangleq 4\widetilde{\mathbb{E}}[x_i x_j] - 1$$

LEMMA 2.3. $\widetilde{\mathbb{E}}[(x_i - x_j)^2] = \frac{1-\rho_{i,j}}{2}$

PROOF. We can compute

$$\begin{aligned} \widetilde{\mathbb{E}}[(x_i - x_j)^2] &= \widetilde{\mathbb{E}}[x_i^2] - 2\widetilde{\mathbb{E}}[x_i x_j] + \widetilde{\mathbb{E}}[x_j^2] \\ &= \widetilde{\mathbb{E}}[x_i] - 2\widetilde{\mathbb{E}}[x_i x_j] + \widetilde{\mathbb{E}}[x_j] = \frac{1 - \rho_{i,j}}{2} \end{aligned}$$

where the second equality comes from the constraint $\widetilde{\mathbb{E}}[x_i^2] = \widetilde{\mathbb{E}}[x_i]$ for all i and the last equality comes from using the fact that $\widetilde{\mathbb{E}}[x_i] = 1/2$. □

Now to complete the analysis, we need to calculate the probability that the edge (i, j) crosses the randomly generated cut $(U, V \setminus U)$:

LEMMA 2.4. $\mathbb{P}[(i, j) \in E(U, V \setminus U)] = \frac{\arccos \rho_{i,j}}{\pi}$

PROOF. First we compute the variance of y_i:

$$\mathrm{Var}(y_i) = \widetilde{\mathbb{E}}\Big[(x_i - \frac{1}{2})^2\Big] = \widetilde{\mathbb{E}}[x_i] - \frac{1}{4} = \frac{1}{4}$$

and similarly

$$\mathrm{Cov}(y_i, y_j) = \frac{\rho_{i,j}}{4}$$

which follows from the way that we have defined Σ and $\rho_{i,j}$. In order for $\arccos \rho_{i,j}$ to be well-defined, we need that $|\rho_{i,j}| \leq 1$. This follows easily by observing that since the variances of y_i and y_j are $1/4$ their covariance must be between $-1/4$ and $+1/4$. Combining these bounds on the covariance with the expression above yields the desired bounds on $\rho_{i,j}$.

Finally we need to compute the probability that our randomized rounding scheme puts i and j on different sides of the cut, which happens exactly when

$$\mathrm{sgn}\left(y_i - \frac{1}{2}\right) \neq \mathrm{sgn}\left(y_j - \frac{1}{2}\right).$$

The trick is to construct the pair of random variables (y_i, y_j) from two independent Gaussian random variables s and t with mean zero and variance one. In particular, (y_i, y_j) has the same distribution as

$$\left(\frac{s}{2} + \frac{1}{2}, \frac{\rho_{i,j} s + \sqrt{1 - \rho_{i,j}^2}\, t}{2} + \frac{1}{2}\right)$$

which implies that

$$\mathbb{P}\Big[\mathrm{sgn}\left(y_i - \frac{1}{2}\right) \neq \mathrm{sgn}\left(y_j - \frac{1}{2}\right)\Big] = \mathbb{P}[\mathrm{sgn}(s) \neq \mathrm{sgn}(\rho_{i,j} s + \sqrt{1 - \rho_{i,j}^2}\, t)].$$

The right hand side is well known to be $\frac{\arccos \rho_{i,j}}{\pi}$, which completes the proof. □

Now we are ready to prove Theorem 2.1:

PROOF. For every edge (i, j) we have

$$\mathbb{P}[(i,j) \in E(U, V \setminus U)] \geq \frac{2 \arccos \rho_{i,j}}{(1-\rho_{i,j})\pi} \widetilde{\mathbb{E}}\Big[(x_i - x_j)^2\Big] \geq \alpha_{GW} \widetilde{\mathbb{E}}\Big[(x_i - x_j)^2\Big].$$

Now, summing over all edges, and by linearity of expectation and pseudo-expectation, we have

$$\mathbb{E}[|E(U, V \setminus U)|] = \sum_{(i,j) \in E} \mathbb{P}[(i,j) \in E(U, V \setminus U)] \geq \alpha_{GW} \widetilde{\mathbb{E}}\Big[\sum_{(i,j) \in E} (x_i - x_j)^2\Big]$$

which completes the proof. □

2.1. Higher Degree Relaxations for MAXCUT. Despite its simplicity, this is the best known approximation algorithm for MAXCUT. In fact any approximation algorithm that achieves a strictly better approximation ratio (independent of the number of nodes) would have spectacular consequences.

In 1994, Arora et al. [**3**,**4**] showed that any proof of length n can be turned into a proof of length polynomial in n in such a way that you can verify its correctness, not by reading it all, but by querying a constant number of bits randomly. These are called probabilistically checkable proofs, and their famous PCP theorem turns out to be the key for showing that various optimization problems are not only NP-hard to solve exactly but even to solve approximately. Over the years, there were many improvements to their construction that had corresponding improvements in hardness of approximation. For example, Håstad [**14**] showed that for any $\epsilon > 0$, it is NP-hard to distinguish between 3-SAT formulas that are perfectly satisfiable and ones where it is possible to satisfy at most a $7/8 + \epsilon$ fraction of the clauses. This is tight in the sense that for any 3-SAT formula, you can always satisfy at least $7/8$ of its clauses.

In 2002, Khot [**18**] formulated a powerful conjecture called the *Unique Games Conjecture*. In some ways, it sounds like a slight modification of the types of NP-hard problems that arise in the PCP literature. The conjecture is still open and its resolution would be a landmark in hardness of approximation (whichever way it works out). Over the years, the community has discovered that this one conjecture leads to an amazing number of tight hardness of approximation results. Most relevant to us, Khot et al. [**19**] and Mossel et al. [**22**] proved:

THEOREM 2.5. *If the Unique Games Conjecture is true then for any $\epsilon > 0$, it is NP-hard to approximate MAXCUT within $\alpha_{GW} + \epsilon$.*

Amazingly, for many of the most natural approaches towards improving on the Goemans-Williamson approximation algorithm and refuting the Unique Games Conjecture, it is not known whether they succeed or fail. We showed how to round the degree 2 Sum-of-Squares relaxation.

OPEN QUESTION 2.6. Does the Sum-of-Squares relaxation for MAXCUT give an approximation algorithm that beats the Goemans-Williamson constant for any constant degree?

In fact, even for degree 4 this is open. In theoretical computer science, the Sum-of-Squares hierarchy plays a unique role: Lower bounds against it are often a source of evidence for a problem being computationally hard. There was a time when there were candidate hard instances of unique games for weaker semidefinite programming hierarchies, and then Barak et al. [**5**] showed that Sum-of-Squares solves all of them. In this way, sometimes the beliefs of the community get swayed by how the story plays out for Sum-of-Squares.

3. Rounding Higher Degree SOS

There are a number of optimization problems where the state-of-the-art algorithms come from Sum-of-Squares. Many of these require degree higher than two. However rounding higher degree Sum-of-Squares relaxations is quite challenging. Our approach of finding a distribution (in our case a Gaussian) that matches the pseudo-moments is doomed to fail for degree three and higher (we will see this as a corollary of the lower bounds later).

3.1. Inequalities Derivable in SOS. So how can we round higher degree Sum-of-Squares relaxations? Most of the rounding algorithms revolve around inequalities that are true for distributions, but can also be proven for pseudo-distributions. Following Barak et al. [**5**] we will show the Cauchy-Schwarz inequality can be derived in Sum-of-Squares, and then we'll use it to give an interesting algorithm for tensor optimization.

The classic Cauchy-Schwarz inequality tells us:

FACT 3.1. If $p(x_1, \cdots, x_n)$ and $q(x_1, \cdots, x_n)$ are polynomials and μ is a distribution on $x_1, \cdots, x_n$ then

$$\mathbb{E}_\mu[p(x)q(x)] \leq \sqrt{\mathbb{E}_\mu[p(x)^2]\mathbb{E}_\mu[q(x)^2]}.$$

We will prove that this inequality is true for pseudo-expectations too:

LEMMA 3.2. *If $p(x_1, \cdots, x_n)$ and $q(x_1, \cdots, x_n)$ are polynomials of degree at most $d/2$ and $\widetilde{\mathbb{E}}$ is a degree d pseudo-expectation then*

$$\widetilde{\mathbb{E}}[p(x)q(x)] \leq \sqrt{\widetilde{\mathbb{E}}[p(x)^2]\widetilde{\mathbb{E}}[q(x)^2]}.$$

Recall that a pseudo-distribution satisfies constraints (1), (2) and (3). Constraint (4) was specific to MAXCUT, or more generally to optimization problems over the hypercube.

PROOF. Suppose that $\widetilde{\mathbb{E}}[p(x)^2], \widetilde{\mathbb{E}}[q(x)^2] > 0$. In this case, by rescaling, we can assume without loss of generality that

$$\widetilde{\mathbb{E}}[p(x)^2] = \widetilde{\mathbb{E}}[q(x)^2] = 1$$

Now using constraint (3) on pseudo-expectations we have $\widetilde{\mathbb{E}}[(p(x) - q(x))^2] \geq 0$ and by expanding we have

$$\widetilde{\mathbb{E}}[p(x)q(x)] \leq \frac{1}{2}(\widetilde{\mathbb{E}}[p(x)^2] + \widetilde{\mathbb{E}}[q(x)^2]) = 1 = \sqrt{\widetilde{\mathbb{E}}[p(x)^2]\widetilde{\mathbb{E}}[q(x)^2]}$$

which completes the proof. □

In general, when you can prove an inequality using a low degree Sum-of-Squares proof then it will hold when you replace expectations by pseudo-expectations. Barak et al. [**5**] showed that you can prove a version of Holder's inequality with Sum-of-Squares. More powerfully, they also showed that you can prove the hypercontractive inequality which implies that for any degree polynomial p of degree at most d

$$\widetilde{\mathbb{E}}[p(x)^4] \leq 9^d(\widetilde{\mathbb{E}}[p(x)^2])^2$$

for any polynomial p of degree d and any degree $4d$ pseudo-expectation. Kauers et al. [**17**] gave a Sum-of-Squares proof for a reverse form of the hypercontractive inequality. De et al. [**11**] gave a Sum-of-Squares proof for the Majority is Stablest Theorem which states that for any Boolean function whose influences are going to zero, majority is the function that has the largest chance of not changing its value when you apply the noise stability operator. Finally, Lei and Sheng [**20**] gave a Sum-of-Squares proof for the powerful Brascamp-Lieb inequality.

3.2. Tensor Optimization. Here we will explain the main ideas in an algorithm of Barak, Kelner and Steurer [**7**] that uses higher degree Sum-of-Squares to approximately optimize the injective norm of a tensor. More precisely, they show:

THEOREM 3.3. *Given an entry-wise nonnegative, symmetric matrix* $M \in \mathbb{R}^{n^2 \times n^2}$, *there is an* $n^{O(\log n/\epsilon^2)}$ *time algorithm to find a unit vector* x *with*

$$\langle x^{\otimes 2}, Mx^{\otimes 2} \rangle \geq \max_{z \ s.t. \ \|z\| \leq 1} \langle z^{\otimes 2}, Mz^{\otimes 2} \rangle - \epsilon \|M\|_2$$

Here $\|M\|_2$ is the spectral norm and is the largest singular value. Even though the result is described as an optimization problem over matrices, you can think of M as coming from an $n \times n \times n \times n$ tensor, in which case we are approximating the quartic form over all unit vectors.

We can once again define the notion of a pseudo-distribution, taking constraints (1), (2) and (3) as usual and instead using

(4′) $\widetilde{\mathbb{E}}[p(x) \sum_{a=1}^n x_a^2] = \widetilde{\mathbb{E}}[p(x)]$

which comes from the polynomial constraint $\sum_{a=1}^n x_a^2 = 1$ specific to the optimization problem at hand.

The basic idea pioneered by Barak, Kelner and Steurer [**7**] is to pretend that the pseudo-distribution is actually a distribution and use its moments to hone in on a feasible solution. And then, if the only inequalities you used are derivable in Sum-of-Squares, reformulate it as a way to round. Essentially all of the later algorithms for utilizing Sum-of-Squares follow this paradigm. It feels almost like a cheat the first time (or few times) you see it.

In this specific tensor optimization problem, we will first construct a unit vector from the pseudo-moments in a simple way. If it works, then great! If not, we will have to find some other way to make progress. In particular suppose that we have found a pseudo-expectation (of some unspecified degree, for now) with objective value γ—i.e.

$$\widetilde{\mathbb{E}}\Big[\sum_{a,b,c,d} M_{a,b,c,d} x_a x_b x_c x_d\Big] = \gamma$$

Here we think of the pair (a, b) as indexing into the rows of M and the pair (c, d) as indexing into the columns. Then we will define a n dimensional vector t with $t_a = \sqrt{\widetilde{\mathbb{E}}[x_a^2]}$. First, we will prove:

CLAIM 3.4. t is a unit vector

PROOF. We can compute

$$\sum_a t_a^2 = \sum_a \widetilde{\mathbb{E}}[x_a^2] = \sum_a \widetilde{\mathbb{E}}[x_a^2] = 1$$

where the second equality comes from constraint (1) and the third comes from constraint (4′). □

Now suppose we get lucky and t satisfies

$$\langle t^{\otimes 2}, Mt^{\otimes 2}\rangle \geq \gamma - \epsilon\|M\|_2$$

Then we are done because

$$\gamma \geq \max_{z \text{ s.t. } \|z\|\leq 1} \langle z^{\otimes 2}, Mz^{\otimes 2}\rangle$$

because the maximum over pseudo-distributions is at least as large as the maximum over unit vectors. The main question is: What do we do if we do not get lucky?

We can define an n^2 dimensional vector s with $s_{a,b} = \sqrt{\widetilde{\mathbb{E}}[x_a^2 x_b^2]}$. We leave it as an exercise to the reader to check that s is also a unit vector.

LEMMA 3.5. *If* $\langle t^{\otimes 2}, Mt^{\otimes 2}\rangle \leq \gamma - \epsilon\|M\|_2$ *then*

$$\|s - t^{\otimes 2}\|_2 \geq \frac{\epsilon}{2}$$

PROOF. First we have

$$\gamma = \widetilde{\mathbb{E}}\Big[\sum_{a,b,c,d} M_{a,b,c,d}x_a x_b x_c x_d\Big] \leq \sum_{a,b,c,d} M_{a,b,c,d}\sqrt{\widetilde{\mathbb{E}}[x_a^2 x_b^2]\widetilde{\mathbb{E}}[x_c^2 x_d^2]} = s^T Ms$$

The inequality above comes from Lemma 3.2. Now rearranging, we have

$$\epsilon\|M\|_2 \leq s^T Ms - \langle t^{\otimes 2}, Mt^{\otimes 2}\rangle = \langle s - t^{\otimes 2}, M(s + t^{\otimes 2})\rangle \leq 2\|M\|_2\|s - t^{\otimes 2}\|_2$$

where the last inequality uses the fact that $\|s + t^{\otimes 2}\|_2 \leq 2$ because both s and t are unit vectors. This completes the proof. □

Now it turns out that we can interpret $\|s - t^{\otimes 2}\|_2$ information theoretically. In particular, we can define two random variables W and W' on $[n] \times [n]$ as follows:

- $\mathbb{P}[W = (a,b)] = \widetilde{\mathbb{E}}[x_a^2 x_b^2] = s_{a,b}^2$
- $\mathbb{P}[W' = (a,b)] = \widetilde{\mathbb{E}}[x_a^2]\widetilde{\mathbb{E}}[x_b^2] = t_a^2 t_b^2$

The important point is that both W and W' have the same marginal distribution on their first (and second) coordinate. By construction, the two coordinates of W' are independent of each other. Now the point is that $\|s - t^{\otimes 2}\|_2$ being lower bounded is the same thing as W and W' being far from each other, in an appropriate choice of distance between random variables. And the only way for them to be far is if the two coordinates of W are somewhat dependent. More precisely we have

$$\|s - t^{\otimes 2}\|_2^2 = \sum_{(a,b)} (\sqrt{\mathbb{P}[W = (a,b)]} - \sqrt{\mathbb{P}[W' = (a,b)]})^2$$

which is (up to a factor of two) exactly the squared Hellinger distance between W and W'. Using various relationships between Hellinger distance and mutual information, it can be shown that:

FACT 3.6. $\|s - t^{\otimes 2}\|_2 \geq \frac{\epsilon}{2}$ implies $I(w_1; w_2) \geq \frac{\epsilon^2}{8}$ where w_1 and w_2 are the two coordinates of W.

$H(\cdot)$ is the entropy function and $H(\cdot|\cdot)$ is the conditional entropy. Moreover

$$I(w_1; w_2) = H(w_1) - H(w_1|w_2) = H(w_2) - H(w_2|w_1)$$

is called the mutual information and it can be thought of as how much the entropy of w_1 reduces (in expectation) when you condition on w_2.

So now we can explain how to complete the rounding procedure. When our first attempt fails, we know that the pseudo-expectation is making coordinates non-negligibly dependent on each other. So we will construct another unit vector using the following two-step procedure:

- Choose i in $[n]$ according to the distribution
$$(\widetilde{\mathbb{E}}[x_1^2], \cdots, \widetilde{\mathbb{E}}[x_n^2])$$
- Define a n dimensional vector u with
$$u_a = \sqrt{\frac{\widetilde{\mathbb{E}}[x_i^2 x_a^2]}{\widetilde{\mathbb{E}}[x_i^2]}}$$

Using the same arguments as before it is easy to show that for any choice of i in the first step, u is a unit vector. Similarly to what we did before, we can use u to define a distribution W'' on $[n]$ where
$$\mathbb{P}(W'' = a) = u_a^2$$
The important point is the following:

FACT 3.7. The distribution of W'' is the same as the distribution of w_1 conditional on $w_2 = i$.

We leave this as an exercise to the reader. Now using what we know about the mutual information between w_1 and w_2, we have that
$$\mathbb{E}_i[H(W'')] = \mathbb{E}_i[H(w_1|w_2 = i)] = H(w_1|w_2) = H(w_1) - I(w_1; w_2) \le \log n - \frac{\epsilon^2}{8}$$
Now our second attempt u could also fail and not achieve objective value at least $\gamma - \epsilon\|M\|_2$. But it turns out there is a way to continue the procedure. We could instead choose a pair $(i,j) \in [n] \times [n]$ according to the distribution
$$(\widetilde{\mathbb{E}}[x_1^2 x_1^2], \cdots, \widetilde{\mathbb{E}}[x_i^2 x_j^2], \cdots, \widetilde{\mathbb{E}}[x_n^2 x_n^2])$$
and define an n dimensional vector u with
$$u_a = \sqrt{\frac{\widetilde{\mathbb{E}}[x_i^2 x_j^2 x_a^2]}{\widetilde{\mathbb{E}}[x_i^2 x_j^2]}}$$
This is yet another candidate unit vector that we can hope achieves large objective value. And if it doesn't, we would continue this process in the obvious way: Next we would condition on a triple (i, j, k) and then on a quadruple and so on. Every time we fail to find a unit vector u that achieves a large enough objective value, we are making progress because there is some distribution on the domain $[n]$ whose entropy is our potential function. When we condition on a larger tuple of variables, we reduce its entropy and this can be quantified through the mutual information. The entropy starts out being at most $\log n$ and it is always nonnegative. So eventually, after at most $\frac{8 \log n}{\epsilon^2}$ rounds, it must terminate.

Now in order to run this procedure, we need our pseudo-expectation to be defined up to degree at least $\frac{16 \log n}{\epsilon^2}$. In the first step, we only used the pseudo-expectation on degree two polynomials. In the second step, we needed it on degree four polynomials, and so on.

This hopefully gives you some sense for how to go about rounding higher degree Sum-of-Squares relaxations. Not only that, but the more inequalities we know how to prove using low degree Sum-of-Squares proofs, the more sophisticated types of reasoning we can utilize when we are working with pseudo-expectations, which in turn yield powerful algorithms that would be quite hard to discover without this machinery in hand.

4. Degree Lower Bounds for Constraint Satisfaction Problems

In the second part of this chapter we will show how prove lower bounds against the Sum-of-Squares hierarchy. Specifically, we will show that it fails to refute random constraint satisfaction problems, even when we let the degree be linear in the number of variables. We will focus on 3-XOR formulas which are defined on variables $x_1, \cdots, x_n$ and have m clauses. Each clause takes three literals (either a variable x_i or its complement $\bar{x}_i$), takes their parity, and checks whether they are equal to some value (that can change from clause to clause). For example we could have

$$x_1 \oplus x_5 \oplus \bar{x}_8 = 1 \text{ and } x_2 \oplus \bar{x}_5 \oplus x_6 = 0 \text{ and so on}$$

For a given assignment, we say that the formula is satisfied if all the clauses are satisfied.

If we are given a 3-XOR formula and want to tell whether there is an assignment that satisfies it, this turns out to be easy. Each clause is an affine constraint in $\mathbb{F}_2$ and so we can rewrite a 3-XOR formula as a system of linear equations over $\mathbb{F}_2$ and use Gaussian elimination to find a satisfying assignment if there is one.

But if instead we want to approximate the largest fraction of clauses that can be simultaneously satisfied, this is NP-hard. More precisely, Håstad [**14**] showed:

THEOREM 4.1. *For any $\epsilon > 0$, it is NP-hard to find an assignment that satisfies a $1/2 + \epsilon$ fraction of the clauses of a 3-XOR formula, even given the promise that these is an assignment that satisfies a $1 - \epsilon$ fraction of them.*

Now assuming $P \neq NP$, we know that Sum-of-Squares also will not be able to approximate the fraction of clauses that can be satisfied within an approximation factor better than $1/2$. A natural question to ask here, and in many other contexts, is whether we can prove such a bound unconditionally.

QUESTION 4.2. Can the Sum-of-Squares hierarchy distinguish between 3-XOR formulas where it is possible to satisfy at least a $1 - \epsilon$ fraction of the clauses, and ones where it is not possible to satisfy a $1/2 + \epsilon$ fraction?

In a seminal work, Grigoriev [**13**] proved that the answer is no. (This result was later rediscovered by Schoenebeck [**23**].) These were the first strong lower bounds against Sum-of-Squares and they hold even when the degree of the relaxation is linear in n. Usually we think of the degree of the relaxation being a constant, or perhaps logarithmic. Then the algorithms that we get out of it run in polynomial or quasi-polynomial time. But when we allow its degree to be linear in n, we are talking about relaxations that we would need exponential time to solve. So Grigoriev's [**13**] lower bound is striking: Sum-of-Squares gives no (additive constant) improvement on the trivial algorithm that can satisfy at least $1/2$ of the clauses until it become essentially a brute-force algorithm that checks all the 2^n possible assignments.

In fact, the lower bound is even stronger for two reasons: (1) It holds even when your goal is to distinguish between 3-XOR formulas where it is possible to satisfy all the clauses, and ones where it is not possible to satisfy a $1/2 + \epsilon$ fraction and (2)

It immediately implies lower bounds for other constraint satisfaction problems like 3-SAT. In regards to (1), what it means is that Sum-of-Squares even fails to solve some problems that are solvable in polynomial time, e.g. by Gaussian elimination. In regards to (2), we can replace a 3-SAT clause of the form, say, $x_1 \vee x_5 \vee \bar{x}_i$ with a 3-XOR clause $x_1 \oplus x_5 \oplus \bar{x}_8 = 1$. When we do this transformation, it is only harder to satisfy the 3-XOR formula in the sense that any assignment that satisfies a clause in the 3-XOR formula will necessarily satisfy the corresponding clause in the 3-SAT formula too. (The converse is not true). Hence if Sum-of-Squares fails to prove that a 3-XOR formula is unsatisfiable, it will fail to prove that the 3-SAT formula is unsatisfiable too.

Before we describe the instances that Grigoriev [**13**] used to fool the Sum-of-Squares hierarchy, first we will perform a helpful transformation. Instead of working with Boolean values we will work with $+1/-1$ values. It is easiest to see this by example. If we start from the clause $x_1 \oplus x_5 \oplus \bar{x}_8 = 1$, we can map it to

$$y_1 \times y_5 \times (-y_8) = -1$$

Now if x_1 takes on the value zero, we assign y_1 the value 1. If x_1 takes on the value 1, we assign y_1 the value -1. It is easy to see that the 3-XOR clause is satisfied if and only if the polynomial constraint on the y_i's is satisfied.

This gives us a way to rewrite the problem of deciding whether a given 3-XOR formula is satisfiable or not as a polynomial optimization problem, to which we can apply our usual trick of turning it into a hierarchy of semidefinite programs that are increasingly tighter relaxations. In particular, we say that $\widetilde{\mathbb{E}}$ is a degree d pseudo-expectation operator for a given 3-XOR formula if

(1) $\widetilde{\mathbb{E}}$ is linear
(2) $\widetilde{\mathbb{E}}[1] = 1$
(3) $\widetilde{\mathbb{E}}[p^2] \geq 0$ for all polynomials p of degree at most $d/2$
(4) $\widetilde{\mathbb{E}}[x_i^2 p] = \widetilde{\mathbb{E}}[p]$ for all polynomials of degree at most $d-2$
(5) $\widetilde{\mathbb{E}}[(x_a x_b x_c - \ell_{a,b,c})p] = 0$ for all polynomials of degree at most $d-3$ and clauses $x_a \times x_b \times x_c = \ell_{a,b,c}$

Grigoriev's [**13**] theorem works for random 3-XOR formulas which are generated as follows: For each clause, we choose three variables (say, x_i, x_j, x_k) uniformly at random. We choose ℓ to be equally likely to be $+1$ or -1. Then we add the constraint

$$x_i \times x_j \times x_k = \ell$$

The important point is:

FACT 4.3. If the number of clauses $m \geq 3n$ then with high probability the formula is unsatisfiable.

This fact is easy to prove by using a Chernoff bound applied to bound the fraction of clauses satisfied by a given assignment and then taking a union bound over all assignments [**4**]. Despite the fact that a random 3-XOR formula is unsatisfiable, we will show that the Sum-of-Squares hierarchy thinks that it is in the following sense:

THEOREM 4.4. [**13**] *For some constants $C \geq 10$ and c, with high probability over the choice of a random 3-XOR formula formula on n variables and $m = Cn$ clauses, there is a degree cn pseudo-distribution that satisfies conditions* (1)–(5)

5. Constructing a Pseudo-Expectation

The main question is: How do we construct a valid pseudo-distribution? Earlier, when we talked about MAXCUT, we used an actual distribution to generate a pseudo-distribution to show that the Sum-of-Squares hierarchy really is a relaxation. But now we must construct something that acts like a distribution on assignments that satisfy every clause, even though there isn't one.

First, we introduce a convenient graph theoretic abstraction:

DEFINITION 5.1. The clause graph is a bipartite graph $G = (U, V, E)$ in which each node $i \in U$ represents a clause, each node $j \in V$ represents a variable and there is an edge (i, j) if and only if the variable x_j is contained in the ith clause.

Also let $\Gamma(S)$ for $S \subset U$ denote the set of nodes in V that have a neighbor in U. The important point is that for a random 3-XOR formula, the constraint graph will be an expander:

DEFINITION 5.2. We say that the clause graph G is left expander with expansion 2 if for every set $S \subset U$ with $|S| \leq n/2$, we have $|\Gamma(S)| \geq 2|S|$.

We will make use of the following fact, that again can be proven by using Chernoff bounds and a union bound:

FACT 5.3. For a random 3-XOR formula on $m \geq 10n$ clauses, with high probability the associated clause graph is a left expander with expansion 2.

Now we are ready to define an iterative procedure for constructing the pseudo-expectation. First, we will focus on defining the pseudo-expectation on monomials where the degree of each variable is at most one. For notational convenience, let

$$x_{\Gamma(i)} = \prod_{j \in \Gamma(i)} x_j$$

which denotes the product of the variables in the ith clause. Also let ℓ_i denote the right hand side of the clause—i.e. what the product of the variables is constrained to be equal to. The intuition is each clause defines a constraint on the joint distribution of the three variables it contains. The pseudo-expectation should respect not only these constraints but also any subsequent constraints we can derive by combining them, provided that when we do so we don't create a constraint defined on more than d variables.

(a) For every clause, set $\widetilde{\mathbb{E}}[x_{\Gamma(i)}] = \ell_i$

(b) For every pair of subsets $S, T \subset [n]$ with $|S \Delta T| \leq d$ set

$$\widetilde{\mathbb{E}}[x_{S \Delta T}] = \widetilde{\mathbb{E}}[x_S]\widetilde{\mathbb{E}}[x_T]$$

where Δ denote taking the symmetric difference.

(c) Whenever rule (b) can no longer be applied, set $\widetilde{\mathbb{E}}[x_S] = 0$ for all remaining S with $|S| \leq d$.

We could encounter a situation where rule (b) reaches a contradiction—i.e. we try to set the pseudo-expectation of some monomial to something different than what we have already set it to. We will show that this happens with vanishing probability.

Now, assuming this process does not fail, we can extend the pseudo-expectation to all degree at most d polynomials. First, if there is a monomial with variables whose degree is larger than one, we can replace any occurrence of x_i^2 with 1. For example we would set

$$\widetilde{\mathbb{E}}[x_1^2 x_2 x_5^3] = \widetilde{\mathbb{E}}[x_2 x_5]$$

We do this because constraint (4) on our pseudo-expectation requires us to. And for any polynomial $p(x) = \sum_S c_S x_S$ we can set

$$\widetilde{\mathbb{E}}[p] = \sum_S c_S \widetilde{\mathbb{E}}[x_S]$$

similarly because constraint (1) on our pseudo-expectation requires us to.

The proof of correctness will rest on the following two lemmas:

LEMMA 5.4. *If the constraint graph is a left expander with expansion* 2 *then the iterative process never fails.*

LEMMA 5.5. *The pseudo-expectation satisfies constraints* (1)–(5)

We will prove the second lemma first. The key ingredient is the following relation

DEFINITION 5.6. For two sets S and T of size at most $d/2$ we say that $S \sim T$ if and only if

$$\widetilde{\mathbb{E}}[x_{S \Delta T}] \neq 0$$

We will prove that it is in fact an equivalence relation:

CLAIM 5.7. $S \sim T$ is an equivalence relation

PROOF. We can directly verify the conditions we need. First, it is symmetric because

$$\widetilde{\mathbb{E}}[x_{S\Delta T}] = \widetilde{\mathbb{E}}[x_{T \Delta S}]$$

Second, it is reflexive because

$$\widetilde{\mathbb{E}}[x_{S \Delta S}] = \widetilde{\mathbb{E}}[x_\emptyset] = \widetilde{\mathbb{E}}[1] = 1$$

Lastly, it is transitive because for any sets S, T and U of size at most $d/2$ we have

$$\widetilde{\mathbb{E}}[x_{S\Delta U}] = \widetilde{\mathbb{E}}[x_{(S\Delta T)\Delta(T \Delta U)}] = \widetilde{\mathbb{E}}[x_{S\Delta T}]\widetilde{\mathbb{E}}[x_{T\Delta U}]$$

where the last equality follows because $|S\Delta T|, |T \Delta U| \leq d$. Now this expression implies that if $S \sim T$ and $T \sim U$ then $S \sim U$. This completes the proof. □

We are now ready to prove Lemma 5.5:

PROOF. Consider a multilinear polynomial p of degree at most $d/2$. By grouping sets $S \subset [n]$ into equivalence classes using the relation $\sim$, we can write

$$p(x) = \sum_{i=1}^{k} \sum_{S \in \mathcal{C}_i} p_S x_S$$

where each $\mathcal{C}_i$ is an equivalence class. Then we can compute

$$\begin{aligned}\widetilde{\mathbb{E}}[p^2] &= \sum_{i=1}^{k}\sum_{j=1}^{k}\sum_{S\in\mathcal{C}_i}\sum_{T\in\mathcal{C}_j} p_S p_T \widetilde{\mathbb{E}}[x_S x_T] \\ &= \sum_{i=1}^{k}\sum_{j=1}^{k}\sum_{S\in\mathcal{C}_i}\sum_{T\in\mathcal{C}_j} p_S p_T \widetilde{\mathbb{E}}[x_{S\Delta T}] \\ &= \sum_{i=1}^{k}\sum_{S\in\mathcal{C}_i}\sum_{T\in\mathcal{C}_i} p_S p_T \widetilde{\mathbb{E}}[x_{S\Delta T}]\end{aligned}$$

The second equality uses the fact that the way we defined our pseudo-expectation operator, we can replace any variable of degree two with the value one. The third equality follow from the fact that if S and T are from different equivalence classes, by definition $\widetilde{\mathbb{E}}[x_{S\Delta T}] = 0$.

Now, for each i, we can choose some $U_i \in \mathcal{C}_i$ arbitrarily. Then we can write

$$\begin{aligned}\widetilde{\mathbb{E}}[p^2] &= \sum_{i=1}^{k}\sum_{S\in\mathcal{C}_i}\sum_{T\in\mathcal{C}_i} p_S p_T \widetilde{\mathbb{E}}[x_{(S\Delta U_i)\Delta(T\Delta U_i)}] \\ &= \sum_{i=1}^{k}\sum_{S\in\mathcal{C}_i}\sum_{T\in\mathcal{C}_i} p_S p_T \widetilde{\mathbb{E}}[x_{(S\Delta U_i)}]\widetilde{\mathbb{E}}[x_{(T\Delta U_i)}] \\ &= \sum_{i=1}^{k}\Big(\sum_{S\in\mathcal{C}_i} p_S \widetilde{\mathbb{E}}[x_{(S\Delta U_i)}]\Big)^2 \geq 0\end{aligned}$$

where the second equality follows from the fact that $|S\Delta U_i|, |T\Delta U_i| \leq d$. Thus our construction satisfies constraint (3) on the pseudo-expectation. It is easy to check the remaining conditions like (1), (2) and (4) directly. For the last condition (5) we have that

$$\widetilde{\mathbb{E}}[(x_{\Gamma(i)} - \ell_i)p] = 0$$

because for every monomial x_S where $\Gamma(i) \subset S$ we set

$$\widetilde{\mathbb{E}}[x_S] = \widetilde{\mathbb{E}}[x_{\Gamma(i)}]\widetilde{\mathbb{E}}[x_{S\setminus\Gamma(i)}]$$

by construction. This completes the proof. □

We will now give a sketch of the proof of Lemma 5.4:

PROOF SKETCH. We can define the notion of a unique neighbor expander: Let $\Gamma_u(S)$ denote the nodes that have exactly one neighbor in the set S. It is easy to prove (but surprising the first time you see it) that if you have a set S of nodes where every node has degree three (like the nodes corresponding to clauses in our clause graph) and $|\Gamma(S)| \geq 2|S|$ then $|\Gamma_u(S)| \geq |S|$.

Now the steps in our iterative procedure for constructing a pseudo-expectation start from clauses and combine them by taking the symmetric difference of their variables. In proof complexity, this is called *resolution*. But whatever set S of clauses we combine in this way, we keep around at least the variables that appear in exactly one clause. These variables prevent us from being able to reach a contradiction because whatever equation we reach, we will always have at least one free variable that makes it possible to satisfy it. □

6. Discussion

With Grigoriev's [**13**] lower bound for random constraint satisfaction problems, we can make good on an assertion we made earlier. When we rounded the degree 2 relaxation for MAXCUT we constructed a Gaussian random variable whose moments match the pseudo-moments of degree one and degree two. Grigoriev's lower bound tells us that this approach is doomed to fail more generally. Recall that $x_S = \prod_{i \in S} x_i$ by convention.

COROLLARY 6.1. *For some constant $c > 0$, there is a pseudo-distribution $\widetilde{\mathbb{E}}$ on variables $x_1, \cdots, x_n$ of degree cn that meets conditions* (1), (2) *and* (3) *but for which no distribution μ satisfies*

$$\mathbb{E}_\mu[x_S] = \widetilde{\mathbb{E}}[x_S]$$

for all $|S| \leq 3$.

PROOF. Consider a random 3-XOR formula and the pseudo-distribution we constructed in our proof of Theorem 4.4. We know that the construction of the pseudo-expectation succeeds with high probability and so, too, the 3-XOR formula is unsatisfiable with high probability. Suppose both events happen, and for the sake of contradiction, suppose there is a distribution μ on assignments whose moments match the pseudo-moments up to degree three. Then from the fact that

$$\mathbb{E}_\mu[x_i^2] = \widetilde{\mathbb{E}}[x_i^2] = 1$$

we conclude that μ is supported on $+1/-1$ valued assignments. Moreover from constraint that for every clause i we have

$$\mathbb{E}_\mu[x_{\Gamma(i)}] = \widetilde{\mathbb{E}}[x_{\Gamma(i)}] = \ell_i$$

we get that μ is supported on satisfying assignments to the 3-XOR formula. But this is impossible because there are no satisfying assignments. □

When we generate a random 3-XOR formula with more clauses (but the same number of variables), it is only easier for Sum-of-Squares to prove that the formula is unsatisfiable. What if we take $m = n^c$? At what value of c does the Sum-of-Squares succeed (with high probability) in proving that the randomly generated formula is indeed unsatisfiable? It turns out that same proof as before (but being more careful keeping track of the various parameters) shows that for $m = n^{3/2-\delta}$ the degree $d = n^{2\delta}$ relaxation fails. This is tight: Barak and Moitra [**8**] showed that when $m \gtrsim n^{3/2}$ the degree six relaxation succeeds and can certify that a random 3-XOR formula is unsatisfiable with high probability.

The way that we constructed a pseudo-expectation may seem magical. But there is actually some useful intuition to why we did what we did. For every monomial $x_{\Gamma(i)}$ that comes from a clause, we had to set $\widetilde{\mathbb{E}}[x_{\Gamma(i)}] = \ell_i$ to satisfy condition (5). Similarly, for every monomial that we could derive from combining clauses, say $S = \Gamma(i)\Delta\Gamma(j)$, we had to set

$$\widetilde{\mathbb{E}}[x_S] = \ell_i \ell_j$$

because we can derive this equation from condition (4) and (5). We leave this as an exercise to the reader. When our iterative procedure terminated (without failing), we are left with monomials whose value we cannot derive. So we made our pseudo-expectation as random as possible and set it to zero. You can think of this as: We have some variable in the monomial whose value we cannot decide on, and if we make it equally likely to be $+1$ or -1, the expectation of the monomial would be zero. The moral[2] is:

> *Make your pseudo-distributions as random as possible, but not randomer.*

As we will discuss next, sometimes this is easier said than done.

6.1. Average-Case Complexity. When we started this chapter, we framed our goal as proving unconditional lower bounds against Sum-of-Squares nominally for problems that we already knew to be NP-hard. However the real reason many people care about Sum-of-Squares lower bounds these days is that they are some of our strongest forms of evidence for various average-case problems actually being hard. Consider the planted clique problem:

DEFINITION 6.2. A random graph G on n nodes is generated as follows: First we include each edge independently with probability $1/2$. Then we choose a set of size k uniformly at random and add in all pairs of edges between these nodes. The goal is to discover which k nodes were chosen.

Information-theoretically, you can find the planted clique with high probability once $k \geq (2+\epsilon)\log n$. But the best known polynomial time algorithms work only for $k \geq c\sqrt{n}$ [**1**]. Making the assumption that there is no polynomial time algorithm that can find $n^{1/2-\delta}$ sized planted cliques turns out to have a number of interesting implications [**10**], however until recently there was no consensus on how likely the assumption is to be true. In fact, it was not even known whether the degree four Sum-of-Squares relaxation does better. Barak et al. [**5**] proved a nearly tight lower bound against Sum-of-Squares:

THEOREM 6.3. *The degree d Sum-of-Squares relaxation for planted clique has value $n^{1/2-o(1)}$ even for degree $d = o(\log n)$.*

Many other problems that appear to be hard on average have been suggested. In this way, Sum-of-Squares can provide a useful way to vet assumptions. It turns out that there are many challenges in proving these types of lower bounds compared to lower bounds against random constraint satisfaction problems. You can think of a pseudo-expectation as a way to answer statistical queries about a hypothetical distribution on a large clique in a graph you're given. Following our usual maxim of making your pseudo-distribution as random as possible, you might think that the probability your pseudo-expectation assigns to every clique of size s (think $s \leq 10$) being in the planted clique ought to be the same [**21**]. This turns out to not work!

The trouble is there is a lot of weak global information. (In contrast, in a random constraint satisfaction problem all of the information was strong and local in the sense that a clause tells you something very particular about the joint distribution of its variables, but this information does not propagate too far.) If some node has a standard deviation higher degree, then it ought to be more likely to be in the planted clique. In fact, it is quite challenging to integrate all of these types of global information in a consistent manner. Finding more general tools to prove lower bounds against Sum-of-Squares in average-case settings is a major challenge going forward [**15**].

[2]This is a cute play on a famous quote by Albert Einstein that we first heard from Boaz Barak.

References

[1] Noga Alon, Michael Krivelevich, and Benny Sudakov, *Finding a large hidden clique in a random graph*, Proceedings of the Eighth International Conference "Random Structures and Algorithms" (Poznan, 1997), Random Structures Algorithms **13** (1998), no. 3-4, 457–466, DOI 10.1002/(SICI)1098-2418(199810/12)13:3/4⟨457::AID-RSA14⟩3.3.CO;2-K. MR1662795

[2] N. Alon and J. Spencer, *The Probabilistic Method*, John Wiley and Sons, 2004.

[3] Sanjeev Arora, Carsten Lund, Rajeev Motwani, Madhu Sudan, and Mario Szegedy, *Proof verification and the hardness of approximation problems*, J. ACM **45** (1998), no. 3, 501–555, DOI 10.1145/278298.278306. MR1639346

[4] Sanjeev Arora and Shmuel Safra, *Probabilistic checking of proofs: a new characterization of NP*, J. ACM **45** (1998), no. 1, 70–122, DOI 10.1145/273865.273901. MR1614328

[5] Boaz Barak, Fernando G. S. L. Brandão, Aram W. Harrow, Jonathan Kelner, David Steurer, and Yuan Zhou, *Hypercontractivity, sum-of-squares proofs, and their applications [extended abstract]*, STOC'12—Proceedings of the 2012 ACM Symposium on Theory of Computing, ACM, New York, 2012, pp. 307–326, DOI 10.1145/2213977.2214006. MR2961513

[6] Boaz Barak, Samuel B. Hopkins, Jonathan Kelner, Pravesh Kothari, Ankur Moitra, and Aaron Potechin, *A nearly tight sum-of-squares lower bound for the planted clique problem*, 57th Annual IEEE Symposium on Foundations of Computer Science—FOCS 2016, IEEE Computer Soc., Los Alamitos, CA, 2016, pp. 428–437. MR3631005

[7] B. Barak, J. Kelner, and D. Steurer, *Rounding Sum-of-Squares relaxations*, Symposium on Theory of Computing, 2014, pages 31–40.

[8] B. Barak and A. Moitra, *Noisy tensor completion via the Sum-of-Squares hierarchy*, Conference on Learning Theory, 2016, pages 417–445.

[9] B. Barak and D. Steurer, *Proofs, beliefs, and algorithms through the lens of sum-of-squares*, `https://www.sumofsquares.org/public/index.html`.

[10] Q. Berthet and P. Rigollet, *Complexity theoretic lower bounds for sparse principal component detection*, Conference on Learning Theory, 2013, pages 1046–1066.

[11] Anindya De, Elchanan Mossel, and Joe Neeman, *Majority is stablest: discrete and SoS*, STOC'13—Proceedings of the 2013 ACM Symposium on Theory of Computing, ACM, New York, 2013, pp. 477–486, DOI 10.1145/2488608.2488668. MR3210809

[12] Michel X. Goemans and David P. Williamson, *Improved approximation algorithms for maximum cut and satisfiability problems using semidefinite programming*, J. Assoc. Comput. Mach. **42** (1995), no. 6, 1115–1145, DOI 10.1145/227683.227684. MR1412228

[13] Dima Grigoriev, *Linear lower bound on degrees of Positivstellensatz calculus proofs for the parity*, Theoret. Comput. Sci. **259** (2001), no. 1-2, 613–622, DOI 10.1016/S0304-3975(00)00157-2. MR1832812

[14] Johan Håstad, *Some optimal inapproximability results*, J. ACM **48** (2001), no. 4, 798–859, DOI 10.1145/502090.502098. MR2144931

[15] Samuel B. Hopkins, Pravesh K. Kothari, Aaron Potechin, Prasad Raghavendra, Tselil Schramm, and David Steurer, *The power of sum-of-squares for detecting hidden structures*, 58th Annual IEEE Symposium on Foundations of Computer Science—FOCS 2017, IEEE Computer Soc., Los Alamitos, CA, 2017, pp. 720–731, DOI 10.1109/FOCS.2017.72. MR3734275

[16] Richard M. Karp, *Reducibility among combinatorial problems*, Complexity of computer computations (Proc. Sympos., IBM Thomas J. Watson Res. Center, Yorktown Heights, N.Y., 1972), Plenum, New York, 1972, pp. 85–103. MR0378476

[17] Manuel Kauers, Ryan O'Donnell, Li-Yang Tan, and Yuan Zhou, *Hypercontractive inequalities via SOS, and the Frankl-Rödl graph*, Proceedings of the Twenty-Fifth Annual ACM-SIAM Symposium on Discrete Algorithms, ACM, New York, 2014, pp. 1644–1658, DOI 10.1137/1.9781611973402.119. MR3376479

[18] Subhash Khot, *On the power of unique 2-prover 1-round games*, Proceedings of the Thirty-Fourth Annual ACM Symposium on Theory of Computing, ACM, New York, 2002, pp. 767–775, DOI 10.1145/509907.510017. MR2121525

[19] Subhash Khot, Guy Kindler, Elchanan Mossel, and Ryan O'Donnell, *Optimal inapproximability results for MAX-CUT and other 2-variable CSPs?*, SIAM J. Comput. **37** (2007), no. 1, 319–357, DOI 10.1137/S0097539705447372. MR2306295

[20] Z. Lei and Y. Sheng, *Sum of Squares proof for Brascamp-Lieb type inequality*, *ArXiv*:1710.01458, 2017.

[21] Raghu Meka, Aaron Potechin, and Avi Wigderson, *Sum-of-squares lower bounds for planted clique [extended abstract]*, STOC'15—Proceedings of the 2015 ACM Symposium on Theory of Computing, ACM, New York, 2015, pp. 87–96. MR3388186

[22] Elchanan Mossel, Ryan O'Donnell, and Krzysztof Oleszkiewicz, *Noise stability of functions with low influences: invariance and optimality*, Ann. of Math. (2) **171** (2010), no. 1, 295–341, DOI 10.4007/annals.2010.171.295. MR2630040

[23] G. Shoenebeck, *Linear level Lasserree lower bounds for certain k-CSPs*, Foundations of Computer Science, 2008, pages 593–602.

Massachusetts Institute of Technology
Email address: moitra@mit.edu

Proceedings of Symposia in Applied Mathematics
Volume **77**, 2020
https://doi.org/10.1090/psapm/077/00685

Applications of sums of squares

Georgina Hall

ABSTRACT. Optimizing over the cone of nonnegative polynomials, and its dual counterpart, optimizing over the space of moments that admit a representing measure, are fundamental problems that appear in many different applications from engineering and computational mathematics to business. In this paper, we review a number of these applications. These include, but are not limited to, problems in control (e.g., formal safety verification), finance (e.g., option pricing), statistics and machine learning (e.g., shape-constrained regression and optimal design), and game theory (e.g., Nash equilibria computation in polynomial games). We then show how sum of squares techniques can be used to tackle these problems, which are hard to solve in general. We conclude by highlighting some directions that could be pursued to further disseminate sum of squares techniques within more applied fields. Among other things, we briefly address the current challenge that scalability represents for optimization problems that involve sum of squares polynomials and discuss recent trends in software development.

A variety of applications in engineering, computational mathematics, and business can be cast as optimization problems over the cone of nonnegative polynomials or the cone of moments admitting a representing measure. For a long while, these problems were thought to be intractable until the advent, in the 2000s, of techniques based on sum of squares (sos) optimization. The goal of this paper is to provide concrete examples—from a wide variety of fields—of settings where these techniques can be used and detailed explanations as to how to use them. The paper is divided into four parts, each corresponding to a broad area of interest: Part 1 covers control and dynamical systems, Part 2 covers probability and measure theory, Part 3 covers statistics and machine learning, and Part 4 covers optimization and game theory. Each part is further subdivided into sections, which correspond to specific problems within the broader area such as, for Part 1, certifying properties of a polynomial dynamical system. Each section is independent of the other sections and is purposefully written so that limited knowledge of the application field is needed. Consequently, a large part of each section is spent on providing a mathematical framework for the field and couching the question of interest as an optimization problem involving nonnegative polynomials and/or moment constraints. A shorter

2010 *Mathematics Subject Classification.* 49-06, 49-02, 90C22, 90C90.

Key words and phrases. Sum of squares optimization, applications, dynamical systems and control, moment problems, option pricing, shape-constrained regression, optimal design, polynomial games, copositive and robust semidefinite optimization.

part explains how to go from these (intractable) problems to (computationally-tractable) sos programs. The conclusion of this paper briefly touches upon some implementation challenges faced by sum of squares optimization and the subsequent research effort developed to counter them.

Part 1. Dynamical systems and control

A dynamical system is a system whose state varies over time. Broadly speaking, a *state* is a vector $x(t) \in \mathbb{R}^n$ that provides enough information on the system at time t for one to predict future values of the state if the system is left to its own devices. For example, if we consider a physical system such as a rolling ball, then one could, e.g., consider the position and instantaneous velocity of its center as a 6-dimensional state vector. As the state vector contains enough information that one can predict its evolution if there is no outside interference, it is possible to relate future states back to the current state via so-called *state equations.* Their expression varies depending on whether the system is *discrete time* or *continuous time.* In a discrete-time system, the state $x(k)$ is defined for discrete times $k = 0, 1, 2, \ldots$, and we have

$$x(k+1) = f(x(k)), \tag{0.1}$$

where f is some function from $\mathbb{R}^n$ to $\mathbb{R}^n$. In a continuous-time system, the state $x(t)$ varies continuously with time $t \geq 0$ and we have

$$\frac{dx(t)}{dt} = f(x(t)), \tag{0.2}$$

where f is again some function from $\mathbb{R}^n$ to $\mathbb{R}^n$. The goal is generally to understand how the *trajectory* $\{x(k)\}_k$, solution to (0.1), or $t \mapsto x(t)$, solution to (0.2), behaves over time. Sometimes such solutions can be computed explicitly and then it is easy to infer their behavior: this is the case for example when f is linear, that is, when $f(x) = Ax$ for some $n \times n$ matrix A; see, e.g., [**11**]. However, when f is more complex, computing closed-form solutions to (0.1) or (0.2) can be hard, even impossible, to do. The goal is then to get insights as to different properties of the trajectories without ever having to explicitly compute them. For example, it may be enough to know that the ball we were considering earlier avoids a certain puddle. This is where sum of squares polynomials come into play—as algebraic certificates of properties of dynamical systems. In Sections 1.1 and 1.2, we will see how we can certify *stability* and *collision avoidance* of polynomial dynamical systems (i.e., dynamical systems as in (0.1) and (0.2) where f is a polynomial) using sum of squares. Other properties such as "invariance" or "reachibility" can be certified using sum of squares as well but are not covered here.

We will also review more complex models than what is given in (0.1) and (0.2). For example, we have assumed here that our dynamical system is *autonomous.* This means that the function f only depends on $x(t)$ or $x(k)$. But this need not be the case. The function f could also depend on, say, an external input $u(t) \in \mathbb{R}^p$. This is a well-studied class of dynamical systems and the vector $u(t)$ is termed a *control.* We will briefly touch upon an example of such a system in Section 1.1. Another alternative to (0.1) and (0.2) could be a direct dependency of f on time on top of its dependency on $x(t)$ or $x(k)$. In this case, such a system is called *time-varying.* We will see an example of such a system in Section 2. In its most general setting, f can be a function of all three: time, state, and control, but we do not cover problems of this type in their full generality here; see, e.g., [**54**] for more information.

1. Certifying properties of a polynomial dynamical system

In this section, we consider a continuous-time polynomial dynamical system:

$$\dot{x} = f(x), \tag{1.1}$$

where $\dot{x}$ is the derivative of $x(t)$ with respect to t and $f : \mathbb{R}^n \to \mathbb{R}^n$ is a vector field, every component of which is a (multivariate) polynomial.

1.1. Stability. Let $\bar{x}$ be an equilibrium point of (1.1), that is a point such that $f(\bar{x}) = 0$. Note that by virtue of the definition, any system that is initialized at its equilibrium point will remain there indefinitely. For convenience, we will assume that the equilibrium point is the origin. This is without loss of generality as we can always bring ourselves back to this case by performing a change of variables $y = x - \bar{x}$ in (1.1). Our goal is to study how the system behaves around its equilibrium point.

DEFINITION 1.1. The equilibrium point $\bar{x} = 0$ of (1.1) is said to be *stable* if, for every $\epsilon > 0$, there exists $\delta(\epsilon) = \delta > 0$ such that $||x(0)|| < \delta \Rightarrow ||x(t)|| < \epsilon,\ \forall t \geq 0$.

Intuitively, this notion of stability corresponds to what we would expect it to be: there always exists a ball around the equilibrium point from which trajectories can start with the guarantee that they will remain close to the equilibrium in the future, where the notion of "close" can be prescribed.

DEFINITION 1.2. The equilibrium point $\bar{x} = 0$ of (1.1) is said to be *locally asymptotically stable* if it is stable and if there exists δ' such that

$$||x(0)|| < \delta' \Rightarrow \lim_{t\to\infty} x(t) = 0.$$

DEFINITION 1.3. The equilibrium point $\bar{x} = 0$ of (1.1) is said to be *globally asymptotically stable* (GAS) if it is stable and if, $\forall x(0) \in \mathbb{R}^n$, $\lim_{t\to\infty} x(t) = 0$.

We will focus on how one can certify *global* asymptotic stability of an equilibrium point in the rest of this section. Analogous results to the ones discussed here exist for both stability and local asymptotic stability and can be found in [**54**, Chapter 4]. The key notion that is used here is that of a *Lyapunov function*, developed by Lyapunov in his thesis [**66**]. The theorem we give below appears in, e.g., [**54**]. It uses the notation $\nabla g(x)$ for the gradient of a function $g : \mathbb{R}^n \to \mathbb{R}$.

THEOREM 1.4. *Let $\bar{x} = 0$ be an equilibrium point for* (1.1). *If there exists a continuously differentiable function $V : \mathbb{R}^n \to \mathbb{R}$ such that*

(i) *V is radially unbounded, i.e., $||x|| \to \infty \Rightarrow V(x) \to \infty$*
(ii) *V is positive definite, i.e., $V(x) > 0, \forall x \neq 0$ and $V(0) = 0$*
(iii) *$\dot{V}(x) := \nabla V(x)^T f(x) < 0$ for all $x \neq 0$ and $\dot{V}(0) = 0$*

then $\bar{x}$ is globally asymptotically stable.

Such a function V is called a Lyapunov function and can be viewed as the generalization of an energy function. The function $\dot{V}$ is the derivative of V with respect to its trajectory as it is equal to $\frac{d}{dt}V(x(t))$ where $x(t)$ is a solution to (1.1). The proof of the theorem is omitted but can be found in [**54**, Chapter 4].

This theorem states a sufficient condition for the equilibrium point to be GAS. Is it the case that whenever the system is GAS, such a Lyapunov function exists? These type of questions give rise to what is known as *converse Lyapunov theorems*. The one given below comes from [**56**] but this precise formulation appears in [**12**].

THEOREM 1.5. *Let f be continuous. If $\bar{x} = 0$ is globally asymptotically stable for* (1.1) *then there exists an ∞-differentiable function $V : \mathbb{R}^n \to \mathbb{R}$ satisfying properties (i)-(iii) of Theorem* 1.4.

Similar theorems to Theorem 1.5 exist for stability and local asymptotic stability; see [**12**]. Theorems such as these do not help us however to explicitly compute a Lyapunov function V, as they do not give a tractable construction of its existence. This is where sum of squares techniques come in useful as we see now. So the techniques can be directly applied, we restrict ourselves to polynomial Lyapunov functions of a certain degree. In practice, this does not seem too restrictive a choice in the context of polynomial dynamical systems and enables a finite parametrization of the Lyapunov functions by their coefficients. A disadvantage however to such a restriction is that, while a C^∞ Lyapunov function was guaranteed to exist for a polynomial dynamical system, a polynomial Lyapunov function is not. In fact, one can find examples of polynomial dynamical systems that do not have polynomial Lyapunov functions; see, e.g., [**7**]. Restricting ourselves to polynomials is not enough however to make the problem tractable. Indeed, searching for polynomial Lyapunov functions that verify (i)-(iii) is a hard problem as it requires constraining polynomials to be positive on $\mathbb{R}^n$. As expected, this is where sum of squares polynomials come into play [**48, 74, 76**].

DEFINITION 1.6. A polynomial function V is a *sum of squares Lyapunov function* for the polynomial system in (1.1) if it vanishes at the origin and satisfies

(i') V is sos
(ii') $-\dot{V}$ is sos.

Note that as $V(0) = 0$ and V is sos, the constant and linear terms of V must be zero. A technical difficulty can arise at this point: constraining V and $-\dot{V}$ to be sos implies that they will be nonnegative, not positive definite as required by Theorem 1.4. Furthermore, V has to be radially unbounded. Though these conditions often hold in practice [**2**, page 41], techniques that guarantee that these constraints hold have also been developed [**75**, Proposition 5].

Replacing (ii)-(iii) by (i')-(ii') comes at a possible loss. It is known for example that if there exists a polynomial Lyapunov function satisfying (i)-(iii) of degree $2d$, it does not follow that an sos Lyapunov function of degree $2d$ exists; see an example in [**9**, Section 3.1]. The related question as to whether an sos Lyapunov function of higher degree exists if a polynomial Lyapunov function exists is open for general polynomial dynamical systems; see [**9**] for some partial results on the topic.

EXAMPLE 1.7. As an illustrative example of what we have seen so far, we consider a model of a jet engine given in [**55**] and revisited in [**19**]. The dynamics of the engine are given by

$$\begin{aligned} \dot{x} &= -y - \frac{3}{2}x^2 - \frac{1}{2}x^3 \\ \dot{y} &= 3x - y. \end{aligned} \tag{1.2}$$

We wish to show that the origin is globally asymptotically stable. Using MATLAB and the software package YALMIP[**64**], we search for a polynomial Lyapunov function V for this system satisfying (i') and (ii'). We start by capping the degree of V at 2, then 4. The solver returns $V = 0$ for degree 2 but a nonzero solution for degree 4. It is easy to check numerically that V is positive definite and radially unbounded, and that $-\dot{V}$ is positive definite too. Hence, the origin is GAS for (1.2). The vector field as well as trajectories of the system (thin lines) and level sets of V (dashed thick lines) are plotted in Figure 1.

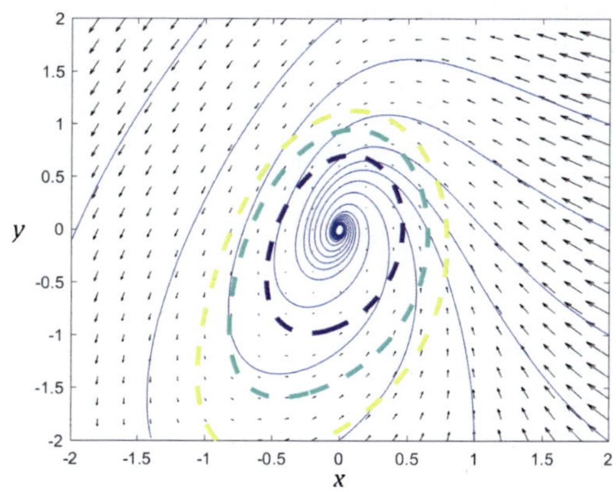

FIGURE 1. Plot of the vector field given in (1.2).

1.1.1. *The specific case of linear systems.* In the particular case where the dynamical system is linear, that is $\dot{x} = Ax$, where A is an $n \times n$ matrix, the previous results simplify considerably. Indeed, $\bar{x} = 0$ is a GAS equilibrium point for the system if and only if a quadratic Lyapunov function V exists. As V is quadratic, it can be parametrized as $V(x) = x^T Px$ where P is a symmetric $n \times n$ matrix. Enforcing conditions (i)-(iii) then simply amounts to searching for a matrix P such that $P \succ 0$ and $A^T P + PA \prec 0$. The search for P is a semidefinite program. If $\bar{x}$ is GAS then such a system will be feasible (see [**24**, Chapter 5] and [**19**, Section 2.2] for the discrete-time case).

1.1.2. *Control.* As mentioned briefly in the introduction, it can be the case that the dynamics depend not only on the state $x(t)$ but also on an external output $u(x(t))$, called a *control*, i.e.

$$\dot{x} = f(x(t), u(x(t))).$$

One can study many properties of controlled systems. In the case of stability, one could wonder for example how to go about designing a controller u in such a way that the size of the region of attraction (i.e., the set of initial states from which a trajectory can start and be asymptotically drawn to its equilibrium) is maximized. If we consider a polynomial control affine system,

$$\dot{x} = f(x) + g(x)u(x),$$

where x is the state variable, $u(x)$ is the control, and f, g are fixed polynomials that are given to us, then this can once more be done using sum of squares optimization. We refer the reader to [**52, 68**] for more information on the topic.

1.2. Collision avoidance. Lyapunov functions originally referred to those functions whose properties certified stability of equilibrium points (such as the ones defined in Theorem 1.4). Now, however, the notion of a Lyapunov function has come to englobe any function that is able to certify properties of a system without requiring explicit computation of its trajectories. Following this broader definition, we will present another category of Lyapunov functions in this subsection, sometimes called *barrier certificates*, which prove that systems are *collision-avoidant.*

We consider again a polynomial dynamical system as in (1.1) and we let $\mathcal{X}_0$ and $\mathcal{X}_u$ to be two sets in $\mathbb{R}^n$. We assume that the trajectories of our system start in $\mathcal{X}_0$, i.e., $x(0) \in \mathcal{X}_0$. We would like to guarantee that all trajectories of (1.1) whose initial states $x(0)$ are in $\mathcal{X}_0$ do not enter the "unsafe region" $\mathcal{X}_u$. Such a system is called collision-avoidant. A sufficient condition for the system to be collision-avoidant is the existence of a barrier certificate, as we describe below.

THEOREM 1.8. [**82**] *Suppose there exists a barrier certificate, namely a continuously differentiable function $B : \mathbb{R}^n \to \mathbb{R}$ that satisfies the following conditions:*

(i) $B(x) > 0$ *for all* $x \in \mathcal{X}_u$
(ii) $B(x) \leq 0$ *for all* $x \in \mathcal{X}_0$
(iii) $\dot{B}(x) = \nabla B(x)^T f(x) \leq 0$ *for all* $x \in \mathbb{R}^n$.

Then there exists no trajectory of (1.1) *that starts from an initial state in* $\mathcal{X}_0$ *and reaches a state in* $\mathcal{X}_u$.

Just as was done previously, we can search for a barrier certificate within the set of polynomial functions. Under the assumption that the sets $\mathcal{X}_u$ and $\mathcal{X}_0$ are closed basic semialgebraic sets, i.e., can be written as the intersection of a finite number of polynomial equalities or inequalities, we can rewrite constraints (i)-(iii) in Theorem 1.8 using sum of squares polynomials.

DEFINITION 1.9. Let $\mathcal{X}_u = \{x \in \mathbb{R}^n \mid g_1(x) \geq 0, \ldots, g_m(x) \geq 0\}$ and $\mathcal{X}_0 = \{x \in \mathbb{R}^n \mid \tilde{g}_1(x) \geq 0, \ldots, \tilde{g}_p(x) \geq 0\}$, where $g_1, \ldots, g_m, \tilde{g}_1, \ldots, \tilde{g}_p$ are polynomials. A sum of squares (sos) barrier certificate is a multivariate polynomial B such that

(i') $B(x) = \epsilon + \sigma_0(x) + \sum_{i=1}^m \sigma_i(x) g_i(x)$, where $\epsilon > 0$ fixed and $\sigma_i, i = 0, \ldots, m$ are sum of squares polynomials
(ii') $-B(x) = \tau_0(x) + \sum_{i=1}^p \tau_i(x) \tilde{g}_i(x)$, where $\tau_i, i = 0, \ldots, p$ are sum of squares polynomials
(iii') $-\dot{B}$ is sos.

Note that searching for such a polynomial is a semidefinite program and that if such a polynomial exists, then it follows that it is a barrier certificate, and hence that the system is collision avoidant.

EXAMPLE 1.10. We illustrate the ideas in this paragraph via an example given in [**82**]. Consider the two dimensional polynomial dynamical system

$$\begin{aligned} \dot{x} &= y \\ \dot{y} &= -x + \frac{1}{3}x^3 - y \end{aligned} \tag{1.3}$$

and the sets $\mathcal{X}_0 = \{(x, y) \mid (x - 1.5)^2 + y^2 \leq 0.25\}$ and $\mathcal{X}_u = \{(x, y) \mid (x + 1)^2 + (y + 1)^2 \leq 0.16\}$. We wish to show that this system is collision avoidant. With this goal in mind, we search for an sos barrier certificate as defined in Definition 1.9 using YALMIP [**64**] and find one of degree 4. In Figure 2, we have plotted the boundaries of the two sets $\mathcal{X}_0$ (in light blue) and $\mathcal{X}_u$ (in black) as well as some trajectories (in dark blue) initialized within $\mathcal{X}_0$ and the 0-level set of our barrier certificate. Note that the 0-level set of our barrier certificate (represented here by the thick green line) can in fact be viewed as a physical barrier.

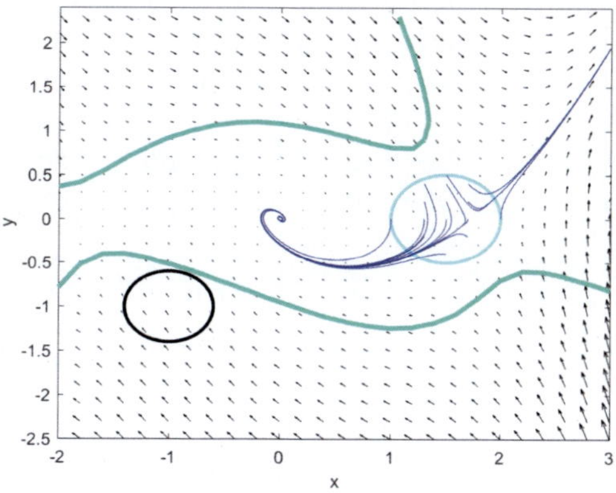

FIGURE 2. The vector field corresponding to the dynamical system in (1.3).

2. Stability of switched linear systems

We now transition from a continuous dynamical system to a discrete dynamical system but with a new twist: in this section, we consider discrete linear systems which are both uncertain and time-varying. More specifically, let $\Sigma := \{A_1, \ldots, A_m\}$ be a set of m real $n \times n$ matrices and let $conv(\Sigma)$ be the convex hull of Σ. Further define the following discrete-time dynamical system

$$x(k+1) = M_k x(k), \text{ where } k = 0, 1, 2 \ldots \text{ and } M_k \in conv(\Sigma). \tag{2.1}$$

This dynamical system is linear, but time-varying as the matrix M_k changes with time, and uncertain as, at every time step, we only know that M_k belongs to the convex hull of a set of fixed matrices, without knowing precisely which one it is. We are interested in knowing whether the equilibrium point $\bar{x} = 0$ is *absolutely asymptotically stable* (AAS) for (2.1), i.e., whether $\lim_{k\to\infty} x(k) = 0$ for any $x(0) \in \mathbb{R}^n$ and any sequence of matrices $\{M_k \in conv(\Sigma)\}_k$. As an example of where such a problem and system may arise, consider, e.g., the task of checking whether a delivery drone is flying in a stable fashion in a windy environment. By linearizing its dynamics around a desired equilibrium point, the behavior of the drone can be modeled locally by a linear dynamical system. However, as this linear dynamical system is unknown due to parameter uncertainty and modeling error, and time-varying due to the effect of the wind, the drone's behavior is better modeled by a system of the type given in (2.1).

Define now, for the same family of m matrices Σ, the following dynamical system, called a *switched linear system*:

$$x(k+1) = A_{\sigma(k)} x(k), \tag{2.2}$$

where $k = 0, 1, 2, \ldots$ is the time index and $\sigma : \mathbb{N} \to \{1, \ldots, m\}$. It so happens that the origin is AAS for (2.1) if and only if it is *asymptotically stable under arbitrary switching* (ASUAS) for (2.2). This means that $\lim_{k\to\infty} x(k) = 0$ for any $x(0) \in \mathbb{R}^n$ and any sequence of matrices $A_{\sigma(1)}, A_{\sigma(2)}, \ldots$ As a consequence, we will study ASUAS of (2.2) but with the idea of applying our results to AAS of (2.1).

2.1. Certifying stability using the Joint Spectral Radius. When $m = 1$, the set Σ is reduced to one matrix A_1 and (2.2) becomes a discrete-time linear system. It is a well-known fact (see, e.g., [**19**, Section 2.2]) that a discrete-time linear system is asymptotically stable if and only if the spectral radius of A_1 is strictly less than one. This can be checked in polynomial-time. When $m \geq 2$, an analogous characterization holds but with a generalization of the notion of spectral radius from one matrix to a family of matrices.

DEFINITION 2.1. [**88**] Let $\Sigma = \{A_1, \ldots, A_m\}$ be a family of m matrices of size $n \times n$. The joint spectral radius (JSR) of Σ is given by

$$\rho(\Sigma) = \lim_{k\to\infty} \max_{\sigma\in\{1,\ldots,m\}^k} ||A_{\sigma(1)} \ldots A_{\sigma(k)}||^{1/k}, \tag{2.3}$$

where $||.||$ is any matrix norm.

Note that when $m = 1$, this definition collapses into $\rho(A_1) = \lim_{k\to\infty} ||A_1^k||^{1/k}$. The right hand side is the spectral radius of A_1 from Gelfand's formula, hence the joint spectral radius is equal to the spectral radius when $m = 1$. As previously mentioned, ASUAS can be characterized using the JSR, which is what we make explicit now.

THEOREM 2.2. [**53**] *The origin is ASUAS for the system given in* (2.2) *if and only if* $\rho(\Sigma) < 1$.

Unlike the linear-system case, where one can decide whether the spectral radius of a matrix is less than one in polynomial time, it is not known whether the problem of testing if $\rho(\Sigma) < 1$ is even decidable (see [**23**] for more computational complexity results relating to the JSR). With the previous result in mind, it comes as no surprise that stability of a switched linear system is not implied by all individual matrices in Σ having spectral radius less than one. For an example of this, consider $\Sigma = \{A_1, A_2\}$ with

$$A_1 = \begin{bmatrix} 0 & 2 \\ 0 & 0 \end{bmatrix} \text{ and } A_2 = \begin{bmatrix} 0 & 0 \\ 2 & 0 \end{bmatrix}.$$

As a consequence, it is of interest to compute upper bounds on the JSR: if these bounds are strictly less than 1, then it will follow that the JSR is as well and the system will be asymptotically stable. A first theorem in this direction, which provides a stepping-stone towards the use of sum of squares polynomials, is given below.

THEOREM 2.3. [**78**, Theorem 2.2] *If there exists a positive definite (homogeneous) polynomial* $p(x)$ *of degree* $2d$ *that satisfies*

$$p(A_i x) \leq \gamma^{2d} p(x), \forall x \in \mathbb{R}^n, \forall i = 1, \ldots, m.$$

Then, $\rho(A_1, \ldots, A_m) \leq \gamma$.

This theorem indicates to us how to use sum of squares polynomials to compute upper bounds on the JSR. We define, as is done in [**78**], the following quantity:

$$\rho_{SOS,2d} := \begin{bmatrix} \inf_{p \text{ of degree } 2d, \gamma} \gamma \\ \text{s.t. } p \text{ sos} \\ \gamma^{2d} p(x) - p(A_i x) \text{ sos}, i = 1, \ldots, m, \\ \int_{S^{n-1}} p(x) dx = 1, \end{bmatrix} \tag{2.4}$$

where S^{n-1} is the hypersphere. Note that for fixed d and fixed γ, the computation of $\rho_{SOS,2d}$ is a semidefinite program. In particular, constraining the integral of $p(x)$ over the hypersphere to be equal to 1 is a linear equation in the coefficients of p; see, e.g., [**38**]. To obtain the smallest γ such that p sos and $\gamma^{2d}p(x) - p(A_i x)$ sos, we proceed by bisection on γ. (Optimizing simultaneously over γ and p results in a nonconvex optimization problem.) We typically fix d and then solve a sequence of semidefinite programs as we bisect over γ. If the optimal value of γ found for that d is satisfactory for our purposes, we stop there; otherwise, we move on to a higher degree. We remark that the accuracy of the bound on the JSR obtained using (2.4) can be quantified (see [**78**, Theorem 3.4]) and finish with an illustrative example of the previously-developed techniques.

EXAMPLE 2.4. Consider a modification of Example 5.4. in [**6**]. We would like to show that the switched linear system defined by the following two matrices

$$A_1 = \frac{1}{\alpha} \begin{bmatrix} -1 & -1 \\ 4 & 0 \end{bmatrix} \text{ and } A_2 = \frac{1}{\alpha} \begin{bmatrix} 3 & 3 \\ -2 & 1 \end{bmatrix},$$

where $\alpha = 3.92$ is stable under arbitrary switching. We are able to show using YALMIP that for $2d = 6$ and $\gamma = 0.9999$, we recover a feasible polynomial p for the SDP given in (2.4). It follows that $\rho(A_1, A_2) \leq 0.9999 < 1$ and hence the system is ASUAS. We showcase this in Figure 3 where we have plotted the 1-level set of p together with three random trajectories of the switched system initalized at the same point. Note that all three trajectories flow towards the origin and remain within the 1-sublevel set of p.

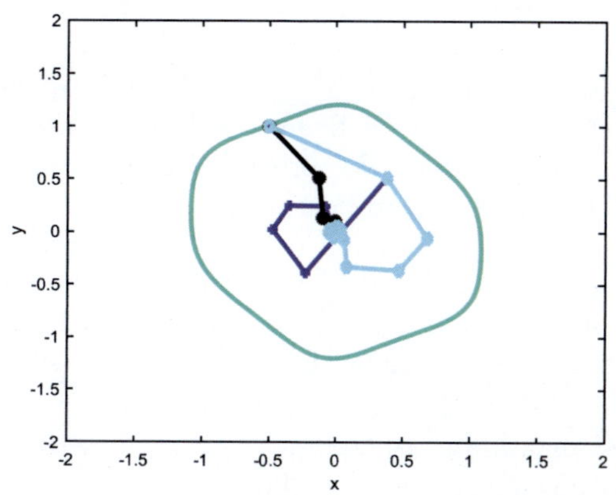

FIGURE 3. Three possible trajectories of the switched linear system described in Example 2.4 together with the 1-sublevel set of the Lyapunov function p obtained (in green)

2.2. Using the dual of (2.4) to generate unstable trajectories. In this subsection, our goal is to generate a sequence of matrices whose asymptotic growth rate is arbitrarily close to the JSR. If the system is unstable, then one can hope to produce unstable trajectories via this method, which would serve to certify its instability. The key component to generate these sequences is duality, specifically deriving the dual of (2.4). We refer the reader to [**62**] for other applications of such techniques as well as a more general version of what is presented here.

Let $\alpha := (\alpha_1, \dots, \alpha_n)^T$ be a vector of nonnegative integers of length n and denote by $|\alpha| := \sum_{i=1}^n \alpha_i$. For a vector of variables $x = (x_1, \dots, x_n)^T$, we can write in shorthand x^α to mean $x_1^{\alpha_1} \dots x_n^{\alpha_n}$. As a consequence, a polynomial $p(x)$ of degree $2d$ in n variables can be written $p(x) = \sum_{|\alpha| \leq 2d} p_\alpha x^\alpha$, where p_α is the coefficient of monomial x^α. There are $N := \binom{n+2d}{2d}$ such coefficients and we denote by $\vec{p}$ the $N \times 1$ vector which contains them. For any given $n \times n$ matrix A, we have $p(Ax) = \sum_{|\alpha| \leq 2d} q_{A,\alpha} x^\alpha$, where $q_{A,\alpha}$ is a linear combination of p_α. Hence, for any $N \times 1$ vector $\tilde{\mu}$, we can define $\tilde{\mu}^A$ such that

$$\langle \tilde{\mu}^A, \vec{p} \rangle = \langle \tilde{\mu}, \vec{q_A} \rangle.$$

Note that each entry of $\tilde{\mu}^A$ is a linear combination of entries of $\tilde{\mu}$. We use the notation $\Sigma^*_{n,2d}$ in the remainder of the subsection for the dual cone of the set of sum of squares polynomials in n variables and of degree $2d$. We refer the reader to Section 4 for a definition of it. For our purposes, it suffices to know that this cone is semidefinite representable. The dual problem of (2.4) can then be written [**62**]

$$\begin{aligned} &\sup_{\tilde{\mu}_1, \dots, \tilde{\mu}_m \in \mathbb{R}^N, \lambda \in \mathbb{R}} \quad \lambda \\ &\text{s.t.} \sum_{i=1}^m \tilde{\mu}_i^{A_i} - \lambda^{2d} \sum_{i=1}^m \tilde{\mu}_i \in \Sigma^*_{n,2d} \\ &\tilde{\mu}_i \in \Sigma^*_{n,2d}, \ \forall i = 1, \dots, m, \\ &\sum_{i=1}^m \langle \tilde{\mu}_i, \vec{s} \rangle = 1, \end{aligned} \tag{2.5}$$

where $\vec{s}$ is the vector of coefficients of $(\sum_{i=1}^n x_i^2)^d$ in the standard monomial basis. Solving this optimization problem for fixed λ and d is a semidefinite program as $\Sigma^*_{n,2d}$ is semidefinite representable. We proceed by bisection again to obtain the largest λ such that the constraints are feasible.

We now describe the algorithm for recovering a sequence of matrices from $\{A_1, \dots, A_m\}$ whose asymptotic growth rate is close to the JSR. To initialize the algorithm, find a feasible solution $(\tilde{\mu}_1^*, \dots, \tilde{\mu}_m^*)$ to (2.5) and pick $\sigma(0) \in \{1, \dots, m\}$ such that $\langle \tilde{\mu}^*_{\sigma(0)}, \vec{s} \rangle > 0$. Such an index is guaranteed to exist given the last constraint of (2.5). Then, for $k = 1, 2, \dots$, pick an index $\sigma(k) \in \{1, \dots, m\}$ such that $\langle \tilde{\mu}^*_{\sigma(k)}, \vec{s}_{A_{\sigma(k)} \cdots A_{\sigma(0)}} \rangle$ is maximum, where $\vec{s}_{A_{\sigma(k)} \cdots A_{\sigma(0)}}$ is the vector of coefficients of $s(A_{\sigma(k)} \cdots A_{\sigma(0)} x)$. Repeat the process. The following guarantee on the growth rate of the sequence thus generated can then be shown.

THEOREM 2.5 (Adapted from Theorem 4.8 in [**62**]). *Consider a family of m $n \times n$ matrices $\{A_1, \dots, A_m\}$. For any positive integer d and for any feasible solution $(\tilde{\mu_1}^*, \dots, \tilde{\mu}_m^*, \lambda^*)$, of (2.5), the algorithm described above produces a sequence*

$\sigma(0), \sigma(1), \sigma(2), \ldots$ *such that*

$$\lim_{k\to\infty} ||A_{\sigma(k)} \ldots A_{\sigma(0)}||_2^{1/k} \geq \frac{\lambda^*}{m^{1/2d}}.$$

2.3. Other areas of application of the JSR. The JSR plays an important role in determining whether a switched linear system is asymptotically stable. But this is far from the only application where it is a relevant quantity. In fact, the concept first started gaining notoriety in the context of the study of wavelets [**20**]. It also appears in economics [**22**], coding theory [**53**], combinatorics on words [**53**], and agent consensus [**21**], to name a few. We give a brief overview of its role in economics here.

In 1973, Wassily Leontief won a Nobel prize in economics for his work on input-output analysis, which models how changes in one sector of the economy can impact other sectors. In his model of inputs and outputs, Leontief divides the economy into n sectors and postulates the following relationship between production and demand: $x = Ax + d$. Here, d is a vector in $\mathbb{R}^n_+$, each component corresponding to demand for the sector i, x is also a vector in $\mathbb{R}^n$, each component describing the production of sector i, and A is a nonnegative $n \times n$ matrix, called the consumption matrix, with entry $(i.j)$ encoding how many units from sector j are required to produce one unit for sector i. The economy is called *productive* if, for any $d \geq 0$, there exists a nonnegative vector x satisfying $x = Ax + d$: this occurs if the spectral radius of A is strictly less than one. However, it can be expected that our knowledge of the consumption matrix is uncertain. It may then be the case that instead of exactly knowing the value of A, we simply know that it belongs to the convex hull of matrices $\{A_1, \ldots, A_m\}$. In this case, to determine whether the economy is productive, one needs to consider the joint spectral radius of $\{A_1, \ldots, A_m\}$ instead; see [**22**] for more details.

3. Related applications

3.1. Fluid dynamics. Fluid dynamics focuses on the study of the flow of fluids such as liquids or gases. Like the problems we considered before, certain properties of this flow can be shown to hold by using Lyapunov analysis. One such property is *stability* of the flow, which we describe now; see [**42, 51**]. Consider a viscous incompressible flow of velocity w and pressure p evolving inside a bounded domain Ω with boundary $\partial\Omega$ under the action of body force f. The velocity and pressure depend on the point x at which we measure them, as well as time t and the way they change is described by the Navier-Stokes and continuity equations:

$$\begin{aligned} \frac{\partial w}{\partial t} + w \cdot \nabla w &= -\nabla p + \frac{1}{R_e}\nabla^2 w + f \\ \nabla \cdot w &= 0, \end{aligned} \tag{3.1}$$

where $\nabla \cdot w$ denotes the divergence of w, $\nabla^2 w$ is the vector Laplacian of w, and R_e is the Reynolds number, which is a constant that depends on the fluid. We further assume that $w = 0$ on the boundary. A steady solution $w = \bar{u}$ and $p = \bar{p}$ to (3.1) (i.e., a solution to (3.1) that does not depend on time) is globally stable if for each $\epsilon > 0$, there exists $\delta > 0$ such that $||w - \bar{u}|| \leq \delta$ at time t_0 implies that $||w - \bar{u}|| \leq \epsilon$ for all $t \geq t_0$. In this case, we say that the flow is not turbulent. A steady flow can be proved to be stable if we are able to construct a Lyapunov function $V(u)$ which is positive definite and decreases on any solution w to (3.1).

The goal is then to obtain the largest Reynolds number R_e such that the flow remains stable. This is traditionally done via an energy stability approach, which means that $V(u)$ is taken to be a very specific Lyapunov function, namely $||u||^2$, where $||u||^2 = \int_\Omega u \cdot u \; d\Omega$. With R_e fixed, determining if the system is stable amounts to solving a linear eigenvalue problem, which is tractable; see [**42**] for more details. However, the value of R_e obtained can be very far from the true value for which the system is unstable. Sum of squares polynomials can then be used to search for improved Lyapunov functions and hence improved values of R_e. We refer the reader to [**42**, **51**] and the references within for additional information on this topic.

3.2. Software verification. The goal of software verification is to ensure that a piece of software satisfies some performance and security specifications. This can include for example finite-time termination, avoiding division by zero, or absence of overflow. A way of doing this is to view the computer program as a discrete-time dynamical system: the program operates over a finite state space and is defined by a transition function, which plays the role of a transition map. From there, it is natural to define analogs to Lyapunov functions (so-called *Lyapunov invariants* in [**87**]), whose purpose it is to certify the aforementioned specifications. We refer the reader to [**87**] for the formal definitions of these concepts and how to use them in practice.

Part 2. Probability and measure theory

In this part, we consider a measure space $(\Omega, \mathcal{F}, \mu)$, where $\Omega \subseteq \mathbb{R}^n$ is the sample space, $\mathcal{F}$ is a σ-algebra over Ω and μ is a measure over $(\Omega, \mathcal{F})$. Recall that a *measure* is a function $\mu : \mathcal{F} \to \mathbb{R}^+ \cup \{+\infty\}$ with two properties: $\mu(\emptyset) = 0$ and $\mu(\cup_{i=1}^{+\infty} F_i) = \sum_{i=1}^{+\infty} \mu(F_i)$ for any pairwise disjoint sets F_i in $\mathcal{F}$. If the measure is a *probability* measure, then it has the additional property that $\mu(\Omega) = 1$. A recurring concept in this Part will be that of *moments* of a measure. Indeed, as we will see in the next section, nonnegative polynomials and moments of a measure are two sides of the same coin via duality. Let $\alpha := (\alpha_1, \ldots, \alpha_n)^T$ be a vector of nonnegative integers of length n and denote by $|\alpha| := \sum_{i=1}^n \alpha_i$. For a vector of variables $x = (x_1, \ldots, x_n)^T$, we can write in shorthand x^α to mean $x_1^{\alpha_1} \ldots x_n^{\alpha_n}$. The moment of order α of a measure μ on $(\Omega, \mathcal{F})$ is then given by $\int_\Omega x^\alpha d\mu(x)$.

In Section 5, we will shift our focus from moments of a measure to moments of a *random variable*. Moments of a random variable X are simply moments of the measure p_X induced by X. (Recall that if X maps from $(\Omega, \mathcal{F}, \mu)$ to $(E, \mathcal{E})$, then $p_X(S) = \mu(\{\omega \in \mathbb{R}^n \mid X(w) \in S\})$ for any set $S \in E$.) The moment of order k of X, where $k \in \{0, \ldots, K\}$, is given by

$$\int_\Omega X^k d\mu = \int_{x \in E} x^k dp_X(x).$$

Note that here k is *not* a multi-index as p_X is a measure over E, a subset of $\mathbb{R}$, and not over Ω, a subset of $\mathbb{R}^n$. By definition of the expectation of a random variable, the moment of order k of X can also be viewed as the expectation $E[X^k]$.

In Section 4, we will focus on a problem called the *moment problem*, and in Section 5, we will work on computing upper bounds on the quantity $p(X \in S)$ where $S \subseteq E$, and the applications of such methods in finance—more specifically, option pricing.

4. The moment problem

In this section, we consider the case where $(\Omega, \mathcal{F}) = (\mathbb{R}^n, \mathcal{B})$, with $\mathcal{B}$ being the Borel σ-algebra over $\mathbb{R}^n$ and focus on the *truncated moment problem*, which is the following inverse problem. Given a a sequence $y := \{y_\alpha\}_{\alpha\in\mathbb{N}^n, |\alpha|\leq 2d}$ where d is some integer, does there exist a *representing measure* μ over $(\mathbb{R}^n, \mathcal{B})$ of y, i.e., a measure such that

$$y_\alpha = \int_\Omega x^\alpha d\mu(x), \forall \alpha \in \mathbb{N}^n, |\alpha| \leq 2d?$$

When there is no constraint on $|\alpha|$ (i.e., the sequence y is infinite), this problem is called the moment problem [**60, 61**]. We follow [**61**] for our presentation of the truncated moment problem.

Let $\mathbb{N}^n_{2d} := \{\alpha \in \mathbb{N}^n \mid |\alpha| \leq 2d\}$ and define

(4.1)

$$\mathcal{M}_{n,2d} := \{\{y_\alpha\}_{\alpha\in\mathbb{N}^n_{2d}} \mid \exists \text{ a measure } \mu \text{ on } \mathbb{R}^n \text{ such that } y_\alpha = \int_{\mathbb{R}^n} x^\alpha d\mu,\ \forall \alpha \in \mathbb{N}^n_{2d}\},$$

i.e., $\mathcal{M}_{n,2d}$ is the set of truncated sequences $\{y_\alpha\}$ for which $\{y_\alpha\}$ has a representing measure μ. It is easy to see that $\mathcal{M}_{n,2d}$ is a convex cone and that the truncated moment problem is exactly the problem of understanding which sequences belong to $\mathcal{M}_{n,2d}$. To answer this question, we consider the dual cone of $M_{n,2d}$.

4.1. Dual cone of $M_{n,2d}$. By definition of a dual cone, we have that

$$(\mathcal{M}_{n,2d})^* = \{\{p_\alpha\}_{\alpha\in\mathbb{N}^n_{2d}} \mid \sum_\alpha p_\alpha y_\alpha \geq 0, \forall \{y_\alpha\} \in \mathcal{M}_{n,2d}\}. \tag{4.2}$$

THEOREM 4.1. *Let $P_{n,2d}$ denote the cone of nonnegative polynomials in n variables and of degree less than or equal to $2d$. We have $P_{n,2d} = (\mathcal{M}_{n,2d})^*$.*

PROOF. Throughout this proof, we will identify a polynomial p in $P_{n,2d}$ by its coefficients p_α in the standard monomial basis, i.e.,

$$P_{n,2d} \triangleq \{\{p_\alpha\}_{\alpha\in\mathbb{N}^n_{2d}} \mid p(x) = \sum_\alpha p_\alpha x^\alpha \geq 0\}.$$

First, $P_{n,2d} \subseteq (\mathcal{M}_{n,2d})^*$. Indeed, if $\{p_\alpha\}_\alpha \in P_{n,2d}$, then for any $\{y_\alpha\}$ in $\mathcal{M}_{n,2d}$, we have

$$\sum_\alpha p_\alpha y_\alpha = \sum_\alpha p_\alpha \int x^\alpha d\mu = \int \sum_\alpha p_\alpha x^\alpha d\mu = \int p(x) d\mu \geq 0$$

as $p(x)$ is nonnegative. Conversely, $P_{n,2d} \supseteq (\mathcal{M}_{n,2d})^*$. Indeed, suppose $p \notin P_{n,2d}$. Then, there exists x_0 such that $p(x_0) < 0$. Let δ_{x_0} be the Dirac measure at point x_0 and let $\{y_\alpha\}_{\alpha\in\mathbb{N}^n_{2d}}$ be the sequence of moments associated to δ_{x_0}. We have

$$\sum_\alpha p_\alpha y_\alpha = \int \sum_\alpha p_\alpha x^\alpha d\delta_{x_0}(x) = \int p(x) d\delta_{x_0}(x) = p(x_0) < 0$$

and hence $\{p_\alpha\} \notin (\mathcal{M}_{n,2d})^*$. □

The corollary below follows.

COROLLARY 4.2. *We have $(P_{n,2d})^* = cl(\mathcal{M}_{n,2d})$, where cl denotes the closure of the set.*

REMARK 4.3. We emphasize that we have identified a polynomial p in $P_{n,2d}$ with its coefficients in the standard monomial basis. (Indeed, the dual cone of $\mathcal{M}_{n,2d}$ should be a subset of $\mathbb{R}^{|\mathbb{N}^n_{2d}|}$.) The interpretation of the dual cone of nonnegative polynomials as sequences of moments is a consequence of having picked a specific basis to represent the polynomials. If that basis changes, e.g., then the interpretation of the dual cone changes as well.

From Theorem 4.1, we are able to conclude that it is hard to establish when a sequence $\{y_\alpha\}$ has a representing measure. Indeed, testing membership to the cone of nonnegative polynomials is a hard task, which implies that testing membership to its dual cone is also hard [**39**]. Theorem 4.1 gives us however a strategy for coming up with necessary conditions for membership to $\mathcal{M}_{n,2d}$. If $\Sigma_{n,2d}$ denotes the cone of sum of squares polynomials of degree $2d$ in n variables, then $\Sigma_{n,2d} \subseteq P_{n,2d}$, which implies that $(P_{n,2d})^* \subseteq (\Sigma_{n,2d})^*$, and hence, $\mathcal{M}_{n,2d} \subseteq (\Sigma_{n,2d})^*$. A necessary condition for membership to $\mathcal{M}_{n,2d}$ is then membership to $(\Sigma_{n,2d})^*$. The latter can be tested using semidefinite programming as we see now.

DEFINITION 4.4. Given a positive integer d, and a truncated sequence $y = (y_\alpha)_{\alpha \in \mathbb{N}^n_{2d}}$, we define the moment matrix $M_d(y)$ to be a symmetric matrix whose rows and columns are indexed by $\alpha \in \mathbb{N}^n_d$ and where, for $\alpha, \beta \in \mathbb{N}^n_d$, the $(\alpha,\beta)^{th}$ entry of the matrix is $y_{\alpha+\beta}$.

THEOREM 4.5. *Let*

$$\mathcal{M}_{\succeq,n,2d} := \{\{y_\alpha\}_{\alpha \in \mathbb{N}^n_{2d}} \mid M_d(y) \succeq 0\}.$$

We have $(\Sigma_{n,2d})^* = \mathcal{M}_{\succeq,n,2d}$.

The proof of this theorem is similar to that of Theorem 4.1. It implies that, given a sequence of $|\mathbb{N}^n_{2d}|$-tuples of numbers y_α, simply checking positive semidefiniteness of the matrix $M_d(y)$ can rule out y_α having a representing measure.

This section can be reworked to take into account measures over arbitrary closed basic semialgebraic sets K. The dual of the set of truncated sequences that have a representing measure μ over K will then simply be the set of polynomials nonnegative over K. Furthermore, one can construct stronger necessary conditions for a representing measure to exist over K (under some assumptions) by simply dualizing well-known hierarchies of inner approximations to $P_{n,2d}$ based on sum of squares; see [**61**] and the references contained there.

REMARK 4.6. As we identified the cone $\Sigma_{n,2d}$ with the set of coefficients of sos polynomials of degree $2d$ and in n variables, $\Sigma_{n,2d}$ was a cone over $\mathbb{R}^{\mathbb{N}^n_{2d}}$. In reality, $\Sigma_{n,2d}$ is a cone over the space of polynomials of degree less than or equal to $2d$, $\mathbb{R}_{2d}[x]$. The dual cone $(\Sigma_{n,2d})^*$ is then the set of linear functionals $L : \mathbb{R}_{2d}[x] \to \mathbb{R}$ such that $L(s) \geq 0$ for any $s \in \Sigma_{n,2d}$. These linear functionals, with the added constraint[1] that $L(1) = 1$, appear under the name *pseudoexpectations* in recent work in the computer science community; see, e.g., [**13**]. The name is due to the fact that, though $L(p) \geq 0$ for any sum of squares polynomial p, it can happen that $L(p) < 0$ for some nonnegative polynomial p. This would never happen for a functional that was truly an expectation.

[1] This latter constraint is because $E[1] = \int d\mu = 1$ for a probability measure.

4.2. The univariate case. The case where $n = 1$ is a noteworthy case for the moment problem (along with the cases $2d = 2$ and $(n = 2, 2d = 4)$) since the set of nonnegative and sum of squares polynomials coincide then. In other words, when $n = 1$, $\Sigma_{1,2d} = \mathcal{P}_{1,2d}$. It then follows from Corollary 4.2 that $(\Sigma_{1,2d})^* = cl(\mathcal{M}_{1,2d})$. This gives rise to the following theorem, the formulation of which is taken from [**19**].

THEOREM 4.7. *Let $y = (y_0, y_1, \ldots, y_{2d})$ be a sequence of real numbers such that $y_0 = 1$. If $y \in \mathcal{M}_{1,2d}$, i.e., if there exists a probability measure μ on $\mathbb{R}$ such that y_i is the ith moment of μ, then $y \in (\Sigma_{1,2d})^*$, i.e.,*

$$M_d(y) = \begin{bmatrix} y_0 & y_1 & y_2 & \cdots & y_d \\ y_1 & y_2 & y_3 & \cdots & y_{d+1} \\ y_2 & y_3 & y_4 & \cdots & y_{2d+2} \\ \vdots & \vdots & \vdots & \ddots & \vdots \\ y_d & y_{d+1} & y_{d+2} & \cdots & y_{2d} \end{bmatrix}$$

is positive semidefinite. Conversely, if $M_d(y) \succ 0$, then y has a representing probability measure μ, i.e., there exists a probability measure μ such that y_i is the ith moment of μ.

Positive definiteness of $M_d(y)$ is needed as one can construct sequences y such that $M_d(y) \succeq 0$ but y does not have a representing measure; see [**19**, Remark 3.147]. The theorem above can also be extended to measures over intervals of $\mathbb{R}$ rather than measures over the whole of $\mathbb{R}$; for this, see again [**19**, Section 3.5.3]. Finally, while this result tells us when a sequence y has a representing measure, it does not directly explain how one should go about constructing such a measure. Some information as to how to do this in practice can be found in [**19**, Section 3.5.5].

4.3. Dual formulation of the polynomial optimization problem. Using the theory developed above, one can view unconstrained polynomial optimization problems (POP) of the type $\min_{x\in\mathbb{R}^n} p(x)$, where p is a polynomial of degree $2d$, or equivalently,

$$\begin{aligned} &\max_{\lambda} \lambda \\ &\text{s.t. } p(x) - \lambda \in P_{n,2d}. \end{aligned} \tag{4.3}$$

as the dual problem of an optimization problem over $M_{n,2d}$. Indeed, one can rewrite the POP as

$$\min_{\text{prob measures } \mu \text{ over } \mathbb{R}^n} \int p(x) d\mu.$$

To see this, let $p^* = \min_{x\in\mathbb{R}^n} p(x)$. Clearly, as $p \geq p^*$, $\int p d\mu \geq p^*$. Conversely, if x^* is a global minimizer of p, then by taking μ to be the Dirac measure δ_{x^*} at x^*, we get $\int p d\delta_{x^*} = p(x^*) = p^*$.

As $p(x)$ is a polynomial of degree $2d$, we have $p(x) = \sum_{\alpha\in\mathbb{N}^n_{2d}} p_\alpha x^\alpha$. Plugging into the previous expression, we get

$$\min_{\text{prob measures } \mu \text{ over } \mathbb{R}^n} \sum_{\alpha\in\mathbb{N}^n_{2d}} p_\alpha \int x^\alpha d\mu.$$

One can then introduce the moments $y_\alpha := \int x^\alpha d\mu$, provided that $\{y_\alpha\}$ has a representing probability measure. The problem becomes:

$$\begin{aligned} &\min_{y_\alpha} \sum_\alpha p_\alpha y_\alpha \\ &\text{s.t. } \{y_\alpha\} \in \mathcal{M}_{n,2d} \text{ and } y_0 = 1. \end{aligned} \tag{4.4}$$

This problem is dual to (4.3) as $P_{n,2d} = (\mathcal{M}_{n,2d})^*$. We refer the reader to [**58**] for more information on this dual formulation. To make the problem tractable, one can replace $\mathcal{M}_{n,2d}$ by its outer approximation $\mathcal{M}_{\succeq,n,2d}$ in (4.4):

$$\begin{aligned} &\min_{y_\alpha} \sum_\alpha p_\alpha y_\alpha \\ &\text{s.t. } \{y_\alpha\} \in \mathcal{M}_{\succeq,n,2d} \text{ and } y_0 = 1. \end{aligned} \tag{4.5}$$

Doing this, we obtain lower bounds on (4.4) via semidefinite programming. This is equivalent to replacing in the dual (4.3) $P_{n,2d}$ by $\Sigma_{n,2d}$:

$$\begin{aligned} &\max_\lambda \lambda \\ &\text{s.t. } p(x) - \lambda \in \Sigma_{n,2d}, \end{aligned} \tag{4.6}$$

which gives us lower bounds on (4.3) using semidefinite programming. The constrained case where $x \in K$, with K being a closed basic semialgebraic set, can also be considered if we work with measures over K instead, as was mentioned earlier.

Dual formulations of the POP can be particularly useful as they provide methods for extracting optimal solutions to the problem, rather than just lower bounds on the optimal value. One way of extracting optimal solutions, presented in [**80**], is to lift the POP into the space of quadratic functions and apply semidefinite programming duality. Another method, that uses the constrained formulation of the dual formulation of the POP presented above can be found in [**49**].

5. Probability bounds given moments of a random variable and applications to option pricing

In this section, we consider a measure space $(\Omega, \mathcal{F}, p)$ where p is a probability measure, a random variable X over $(E, \mathcal{E})$ where E is a subset of $\mathbb{R}$ and $\mathcal{E}$ is a σ-algebra over E, and its induced measure p_X. While the results given here apply to random variables, they can easily be extended to random vectors. Let $\{y_k\}_{k\in\{0,\dots,K\}}$ be a sequence of scalars and let Φ be the set of probability measures

$$\Phi := \{p_X \mid \int_E x^k dp_X(x) = y_k, \forall k \in \{0,\dots,K\}\}. \tag{5.1}$$

The set Φ can also be identified with the set of random variables X whose order-k moments coincide with y_k for $k \in \{0,\dots,K\}$. We use the notation Φ for both sets and assume that Φ is non-empty (or in other words, $\{y_k\}$ always has at least one representing measure). As we shall see, the problems we consider in Sections 5.1 and 5.2 involve optimizing some functional of p_X over the set Φ.

5.1. Probability bounds given moments of a random variable. Given a sequence $\{y_k\}_{k\in\{0,\dots,K\}}$ of moments of X, and a set $S \subseteq E$ described by polynomial inequalities, we wish to derive a "tight" upper bound on $p_X(S) =: p(X \in S)$, i.e., derive $\sup_{p_X\in\Phi} p_X(S) = \sup_{X\in\Phi} p(X \in S)$. This problem is a seminal problem in probability theory and two of the most ubiquitous inequalities in the field, namely that of Markov and Chebychev, exactly cover problems of this type. The Markov inequality states that, for any nonnegative random variable X and positive scalar a,

$$p(X \geq a) \leq \frac{E[X]}{a}. \tag{5.2}$$

Similarly, the Chebychev inequality states that for any random variable X,

$$p(|X - E[X]| \geq t) \leq \frac{var(X)}{t^2}. \tag{5.3}$$

Note that the upper bounds on the probabilities in both inequalities only depend on the distribution of the random variable X via its moments.

How can we tackle the more general problem stated above? It is straightforward to see that the problem $\sup_{p_X \in \Phi} p_X(S)$ can be formulated as

$$\begin{aligned} &\sup_{p_X} \int_E \mathbf{1}_S dp_X \\ &\text{s.t. } \int x^k dp_X(x) = y_k, \forall k \in \{0, \ldots, K\}, \end{aligned}$$

where $\mathbf{1}_S$ refers to the indicator function of S, i.e., a function that takes value 1 on S and 0 elsewhere. The dual to this problem reads

$$\begin{aligned} &\inf_{\lambda_k} \sum_{k=0}^{K} \lambda_k y_k \\ &\text{s.t. } \sum_{k=0}^{K} \lambda_k x^k \geq \mathbf{1}_S, \forall x \in E. \end{aligned} \tag{5.4}$$

with strong duality holding under mild conditions; see, e.g., [**17**]. If we assume that E and S are basic semialgebraic sets and if we replace the constraint that $\sum_{k=0}^{K} \lambda_k x^k \geq \mathbf{1}_S, \forall x \in E$ by the two constraints $\lambda(x) - 1 \geq 0, \forall x \in S$ and $\lambda(x) \geq 0, \forall x \in E$, then this problem can be tackled using sum of squares polynomials. Indeed, one can enforce nonnegativity of the polynomials $\lambda(x) - 1$ and $\lambda(x)$ over E and S respectively by using sum of squares certificates of nonnegativity. In our case, i.e., the case where X is a random variable, this can be done with no loss [**19**]. For more complex cases such as the multivariate case (i.e., X is a random vector), we refer the reader to [**17, 59**].

EXAMPLE 5.1. Let X be a nonnegative random variable whose distribution is unknown but its first moment $E[X]$ is known. We would like to find an upper bound on $p(X \geq a)$ for a given scalar $a > 0$. Following our previous notation, we have $K = 1$, S is $[a, \infty)$, and E, which is where X takes its values, is $[0, \infty)$. The fact that $K = 1$ implies that $\lambda(x)$ is an affine polynomial, i.e., $\lambda(x) = \lambda_0 + \lambda_1 x$. Hence the problem to solve is the following:

$$\begin{aligned} &\min_{\lambda} \lambda_0 + \lambda_1 E[X] \\ &\text{s.t. } \lambda(x) - 1 \geq 0, \ \forall x \geq a \\ &\qquad \lambda(x) \geq 0, \ \forall x \geq 0. \end{aligned}$$

(Note that $y_0 = 1$ as we are considering a probability measure.) One can rewrite the constraints exactly (see [**19**, Section 3.3.1]) to obtain a linear program. Picking

$\lambda(x) = \frac{1}{a}x$, it is straightforward to see that $p(X \geq a) \leq E[X]/a$, which is Markov's inequality (5.2). By appropriately choosing a random variable X, it can further be shown that this upper bound is tight if $E[X]/a \leq 1$. If $E[X]/a \geq 1$, then a tight upper bound is given by 1.

EXAMPLE 5.2. Let X be a random variable whose distribution is unknown but whose first and second order moments, $E[X]$ and $E[X^2]$, are known. Given a scalar t, we would like an upper bound on $p(|X - E[X]| \geq t)$ that involves only $E[X]$ and $E[X^2]$. In this case, $K = 2$, $S = (-\infty, -t + E[X]] \cup [t + E[X], +\infty)$, $E = \mathbb{R}$, and we have $\lambda(x) = \lambda_0 x + \lambda_1 x + \lambda_2 x^2$. The problem can then be written as

$$\begin{aligned} &\min_{\lambda} \lambda_0 + \lambda_1 E[X] + \lambda_2 E[X^2] \\ &\text{s.t. } \lambda(x) - 1 \geq 0, \ \forall x \in (-\infty, -t + E[X]] \cup [t + E[X], +\infty) \\ &\qquad \lambda(x) \geq 0, \forall x \in \mathbb{R}. \end{aligned} \tag{5.5}$$

This problem can be recast as a semidefinite program; see, e.g., [**19**, Theorem 3.72] and solved. A feasible solution to the problem is $\lambda(x) = \left(\frac{x-E[X]}{t}\right)^2$ and its corresponding optimal value is $var(X)/t^2$. It follows that $p(|X - E[X]| \geq t) \leq var(X)/t^2$, which is Chebychev's inequality (5.3). By appropriately choosing a random variable X, it can be shown that this upper bound is tight when $var(X)/t^2 \leq 1$. When $var(X)/t^2 \geq 1$, 1 is a tight upper bound.

5.2. Applications to option pricing. Let X be the (random) price of an asset and p_X its probability distribution. Though p_X is unknown, we assume that the first and second order moments of X, denoted by y_1 and y_2, are known (e.g., estimated from past data). The zero-th order moment of X is trivially $y_0 = 1$. We now consider a European call option on the asset with strike price k. This is a derivative security which gives the buyer of the call two options on the day it expires: either (s)he buys a fixed amount of the asset at price k, or (s)he does nothing. Hence, the payoff of the buyer of the option will be $\max(0, X - k)$ where X is the price of the asset on the day the call expires: indeed, if the price of the asset is greater than k, then the buyer will use his or her option to get it at the reduced price of k, thus making $X - k$; if the price of the asset is less than k however, then the buyer will chose to not use his or her option, thus making 0. A fair price for this option would be $E_{p_X}[\max\{0, X - k\}]$, where the expectation is taken with respect to the unknown probability distribution of X. Note that with such a price, the seller does not make a profit on average, but simply breaks even. However, to hedge against uncertainty in the distribution of X, the seller choses to pick

$$\sup_{p_X \in \Phi} E_{p(X)}[\max\{0, X - k\}]$$

where Φ is as in (5.1) with $E = [0, +\infty)$ (the price of the asset is always nonnegative) and $K = 2$. As discussed before, the problem above can be formulated as:

$$\begin{aligned} &\max_{p_X} \int_{\mathbb{R}+} \max\{0, x - k\} dp_X(x) \\ &\text{s.t. } \int_{\mathbb{R}+} x^k dp_X(x) = y_i, i = 0, 1, 2. \end{aligned}$$

The dual to this problem is

$$\begin{aligned}&\min_{\lambda_k} \sum_{k=0}^{2} \lambda_k y_k\\&\text{s.t. } \sum_{k=0}^{2} \lambda_k x^k \geq \max\{0, x-k\}, \forall x \in \mathbb{R}^+.\end{aligned}$$

The constraint can equivalently be reformulated as $\sum_{k=0}^2 \lambda_k x^k \geq 0, \forall x \in [0,k]$ and $\sum_{k=0}^2 \lambda_k x^k \geq x-k, \forall x \in [k,+\infty)$, and the problem can then be solved using semidefinite programming. We refer the interested reader to [**16**] for other examples of problems of this type. Other areas where optimal bounds on probabilities of events can be useful are decision analysis [**92**] and queuing theory [**99**].

Part 3. Statistics and machine learning

Both of the sections in this part revolve around *regression.* Regression is one of the most fundamental problems in statistics, with applications in many different areas, including the social and physical sciences. We briefly review the problem here. Let $\{(x_i, y_i)\}_{i=1,\ldots,m}$ be a series of data points with $x_i \in \mathbb{R}^n$ being a feature vector, and $y_i \in \mathbb{R}$ being the output variable. We denote by x_i^j the jth component of the vector x_i. It is assumed that there is a relationship between x_i and y_i of the form

$$y_i = f(x_i) + \epsilon_i,\ i = 1, \ldots, m$$

where ϵ_i is some random noise with $E[\epsilon_i] = 0$, finite variance $var(\epsilon_i) = \sigma^2 < \infty$, and ϵ_i independent from ϵ_j. The goal of regression is to find a function f (called a *regressor*) within a class of functions $\mathcal{F}$ such that the error between $f(x_i)$ and y_i is minimized. The notion of error can be, e.g., that of least squares error, which gives us the problem

$$\min_{f \in \mathcal{F}} \sum_{i=1}^{m} (y_i - f(x_i))^2. \tag{0.3}$$

When $\mathcal{F}$ contains functions that are completely described by a set of parameters $\theta \in \mathbb{R}^p$, the regression is called *parametric* and the optimization can be done over the parameters instead of over $\mathcal{F}$. The case where $\mathcal{F} = \{f \mid f(z) = \theta_0 + \theta_1 z_1 + \ldots + \theta_n z_n$, where $\theta_0, \ldots, \theta_n \in \mathbb{R}\}$, for example, is linear regression and one can f by solving an unconstrained convex quadratic program. The case we consider is *polynomial* regression, i.e.,

$$\mathcal{F} = \{f \mid f(z) = \sum_{\alpha \in \mathbb{N}_d^n} \theta_\alpha z^\alpha\}, \tag{0.4}$$

where, as in Section 4, we let $\mathbb{N}_d^n = \{\alpha \in \mathbb{N}^n \mid \alpha_1 + \ldots + \alpha_n \leq d\}$ and, for $z = (z_1, \ldots, z_n)^T$, we use the shorthand z^α to mean $z_1^{\alpha_1} \ldots z_n^{\alpha_n}$. In Section 6, we will show how sum of squares polynomials can be used to enforce shape constraints on the regressor f. In Section 7, we will see how we can use sum of squares techniques to optimally pick the feature vectors x_i.

6. Shape-constrained regression

In this section, we consider the regression problem described above and assume that the feature vectors x_i belong to a full-dimensional box B in $\mathbb{R}^n$. Our goal is to fit a function f to the data that minimizes the least squares error within a class of functions $\mathcal{F}$ that have a specific shape (e.g., convex over the box B or monotonous over B in certain directions). We call this problem *shape-constrained* regression. Shape-constrained regression is a very natural problem to consider. In economics for example, if one wants to model a utility function by fitting a regressor to data, then it would make sense to enforce concavity of the regressor. Likewise, we can readily imagine that a number of outputs would depend monotonically on inputs (think, e.g., of the Body Mass Index (BMI) of a person with respect to his or her calorie intake, or the quantity of honey produced in a hive as a function of number of bees). Because of its omnipresence, there have been a number of methods developed to address this problem; see [**44**–**46**, **63**, **69**, **90**]. Here, we consider a method that relies on sum of squares programming, developed in, e.g., [**3**, **67**]. One of its main advantages is that it scales polynomially in the number of features of the problem, which is often a caveat in other methods. We discuss it in more depth below.

Let's consider first the case where we would like to enforce monotonicity of our regressor over B with respect to component j, i.e., we want

$$z_j \mapsto f(z_1, \ldots, z_{j-1}, z_j, z_{j+1}, \ldots, z_n)$$

to be increasing for all $(z_1, \ldots, z_{j-1}, z_{j+1}, \ldots, z_n)$ in the appropriate domain. We assume here that f is continuously differentiable. This is then equivalent to imposing that

$$\frac{\partial f(z)}{\partial z_j} \geq 0, \ \forall z \in B.$$

If $\rho \in \mathbb{R}^n$ is a vector that encodes the monotonicity profile of f with respect to each one of its variables, i.e., $\rho_j = 1$ (resp. $0, -1$) if f is increasing (resp. non-monotonic, decreasing) with respect to component j, then the monotonicity-constrained regression problem can be written:

$$\begin{aligned} \min_f \ & \sum_{i=1}^m (y_i - f(x_i))^2 \\ \text{s.t.} \ & \rho_j \frac{\partial f(z)}{\partial z_j} \geq 0, \ \forall z \in B. \end{aligned}$$

To make the problem amenable to computation, we let $\mathcal{F}$ be as in (0.4), i.e., f is assumed to be a polynomial of degreee d. The problem remains hard to solve however because of the nonnegativity constraint over the box. Indeed, one can show that even testing whether a polynomial f of degree d has monotonicity profile ρ, over a box B is NP-hard, for d as low as 3 [**3**]. We consequently replace the nonnegativity constraint by a constraint that involves sum of squares polynomials—see [**19**, Section 3.4.4] for different ways to do this—and the problem becomes a semidefinite program. The theorem below gives an idea as to the quality of these successive approximations.

THEOREM 6.1. [**3**] *Let f be a continuously differentiable function with a given monotonicity profile ρ over B. For any $\epsilon > 0$, there exists an integer d and a* *polynomial p of degree d such that*

$$\max_{x \in B} |f(x) - p(x)| < \epsilon$$

and such that p has same monotonicity profile ρ over B. Furthermore, this monotonicity profile can be certified using a sum of squares certificate.

Let us consider now the case where we would like to enforce convexity of our regressor f over B. We assume that f is twice continuously differentiable and that H_f denotes the Hessian of f. This is then equivalent to imposing $H_f(y) \succeq 0, \forall y \in B$, which is in turn equivalent to $z^T H_f(y) z \geq 0,\ \forall z \in \mathbb{R}^n,\ \forall y \in B$. Hence the convexity-constrained regression problem can be written

$$\begin{aligned} &\min_f \sum_{i=1}^m (y_i - f(x_i))^2 \\ &\text{s.t. } z^T H_f(y) z \geq 0,\ \forall z \in \mathbb{R}^n,\ \forall y \in B. \end{aligned}$$

We follow the same scheme as previously: we restrict ourselves to polynomial functions, and then replace the nonnegativity constraint of the polynomial (in z and y) $z^T H_f(y) z$ by a constraint that involves sum of squares polynomials. Indeed, as before, the problem of testing whether a polynomial of degree d is convex over a box is NP-hard, even for $d = 3$ [**4**]. One can show an analogous result to Theorem 6.1 for convexity-constrained regression. Note that if we have the appropriate prior information, both types of constraints (monotonicity and convexity) can be combined.

REMARK 6.2. Let f be a polynomial in n variables. If $y^T H_f(x) y$ is constrained to be a sum of squares (as a polynomial in x and y) then f is said to be *sos-convex*. This is a sufficient condition for (global) convexity and optimizing over the set of sos-convex polynomials is a semidefinite program; see [**10**] for more information on the concept of sos-convexity. The above example uses a variant of sos-convexity as we wish to find sufficient conditions for convexity *over a box*. Sos-convexity is a notion that can be used in many other applications. For example, some problems in computer vision require fitting a convex shape to a cloud of $3D$ data points in such a way that the volume of this shape is as small as possible: this can be done e.g. by requiring that these points belong to the sublevel set of an sos-convex polynomial; see [**5, 67**].

EXAMPLE 6.3. We now give an example, taken from [**3**], relating to the prediction of weekly wages from past data. The data used comes from the 1988 Current Population Survey and is freely available under the name ex1029 in the Sleuth2 R package [**84**]. It contains 25361 observations and 2 numerical features: years of experience and years of education. We expect wages to increase with respect to years of education and be concave with respect to years of experience. We run both an unconstrained polynomial regression (denoted by UPR), i.e., $\mathcal{F}$ is the set of polynomials of a certain degree in (0.2), and a convexity-constrained and monotonocity-constrained regression (denoted by Hybrid and described above) on the data. This is done by computing the Root Mean Squared Error (RMSE) for the data with 10-fold cross validation. The results are given in Figure 4 with varying degrees of the polynomial regressor. Note that for the training data, obviously UPR performs better than Hybrid as it is less constrained and can overfit. The Hybrid method however has a much better generalization error than UPR.

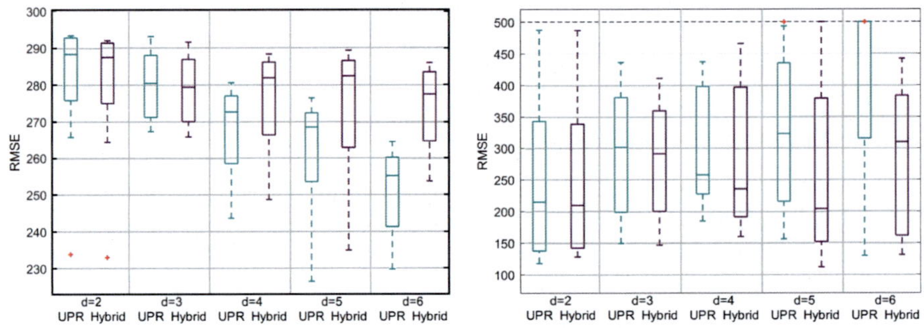

(a) RMSE on training data (b) RMSE on testing data

FIGURE 4. Comparative performance of UPR and Hybrid on testing and training sets.

7. Optimal design

We once again consider a regression setting, but this time we are interested in the problem of generating the input data, $\{(x_i, y_i)\}_{i=1,\dots,m}$. Statisticians make a difference between the case where the person conducting the study can choose the feature vectors $\{x_i\}_i$, and the case where the feature vectors $\{x_i\}_i$ are imposed. The latter case is called an *observational study*. An illustrative example would be that of a study of the impact of the amount of cigarettes smoked on the development of lung cancer: our data would contain the amount of cigarettes that each participant chooses to smoke and whether he or she develops lung cancer. Ethically, it would not be possible for us to ask our participants to smoke more to improve our input data. Of interest in this section is the other case, namely the case where the x_i can be fixed to certain values by the experimenter. This is called an *experimental study*. As an example of such a study, consider the problem of measuring the degree of corrosion of steel under the effects of humidity and temperature. By placing a piece of steel in an environment controlled for humidity and temperature, one is able to obtain the degree of corrosion for any values of humidity and temperature that one wishes to have. This set-up is particularly interesting to statisticians as it enables the experimenter to choose advantageous values of the features. The process of choosing such values is termed *experimental design*. In this section, we will be interested in using sum of squares techniques to understand how to design experiments in an optimal way. We follow the presentation given in [**31**].

7.1. Mathematical formulation of the optimal design problem. Consider the regression set-up described at the beginning of Part 3. We further assume that $m \geq \binom{n+d}{n}$ and that the vectors $\{x_i\}$ can be picked within a compact set $\mathcal{X} \subset \mathbb{R}^n$, described by a finite number of polynomial inequalities. Note that it may be the case that the optimal way of picking the vectors $\{x_i\}$ involves repeating the same value twice, e.g., $x_1 = x_2$. As a consequence, our goal is to choose l ($l \leq m$) distinct values, $t_1, \dots, t_l \in \mathcal{X}$ that the $\{x_i\}$ take, together with the number of times n_k that value t_k is taken. If we let $w_k = \frac{n_k}{m}$, this information can be summarized in a $(n+1) \times l$ *design matrix*

$$\xi = \begin{bmatrix} t_1 & \dots & t_l \\ w_1 & \dots & w_l \end{bmatrix}, \text{ where } w_k = \frac{n_k}{m}, \tag{7.1}$$

which is what we would like to obtain at the end of the optimization process. In the rest of this section, for convenience, we denote the vector of standard monomials of degree up to d in n variables by $z_d(x) = (1, x_1, x_2, \dots, x_n, \dots, x_n^d)$, and by θ the corresponding vector of coefficients so that $\sum_{\alpha \in \mathbb{N}_d^n} \theta_\alpha x_i^\alpha = \theta^T z_d(x)$. For ease of exposition, we also assume, until otherwise mentioned, that $l = m$ (i.e., the feature vectors x_i are distinct), and that $\sum_{k=1}^m z_d(t_k) z_d(t_k)^T \succ 0$.

What should be the objective when picking ξ? This depends on what we would like to achieve. In our case, assuming our estimator for θ is the least squares estimator

$$\hat{\theta} = \arg\min_\theta \sum_{k=1}^m (y_k - \theta^T z_d(t_k))^2, \tag{7.2}$$

it may be of interest to minimize, in some sense, the variance of $\hat{\theta}$. Indeed, under the assumptions we have on ϵ_i, $\hat{\theta}$ is an *unbiased* estimator of θ, which means that $E[\hat{\theta}] = \theta$; i.e., on average both quantities are equal. It may then be of interest to ask that $\hat{\theta}$ deviate as little as possible from θ overall: this is exactly equivalent to minimizing the variance of $\hat{\theta}$. Of course, the variance of $\hat{\theta}$ is a matrix here as $\hat{\theta}$ is a vector, so when we claim to minimize the variance of $\hat{\theta}$, we actually mean minimizing the 2-norm of its covariance matrix, or some other measure.

Consequently, our goal is to compute the covariance matrix of θ. Under our assumptions, the objective function in (7.2) is a strictly convex quadratic function and hence has a unique minimum. Using the first-order necessary condition for optimality, we obtain

$$\hat{\theta} = \sum_{k=1}^m y_k \left(\sum_{k=1}^m z_d(t_k) z_d(t_k)^T \right)^{-1} z(t_k).$$

Replacing $y_k = \theta^T z_d(x_k) + \epsilon_k$ in the previous expression, we get

$$\hat{\theta} = \theta + \left(\sum_{k=1}^m z_d(t_k) z_d(t_k)^T \right)^{-1} \sum_{k=1}^m \epsilon_k z_d(t_k).$$

(This confirms that $\hat{\theta}$ is an unbiased estimator of θ as previously claimed.) Using standard covariance and variances identities as well as the properties of ϵ_i, it follows that the variance matrix of $\hat{\theta}$ is given by $\Sigma = \sigma^2 \left(\sum_{k=1}^m z_d(t_k) z_d(t_k)^T \right)^{-1}$. We write $\Sigma = \Sigma(\xi)$ as one can see that the covariance matrix of $\hat{\theta}$ only depends on ξ and the variance of ϵ_i, σ^2, which we assume to be a fixed constant. In the more general case where the points t_k are not assumed distinct, the covariance matrix is given by

$$\Sigma(\xi) = \sigma^2 \left(\sum_{k=1}^l n_k z_d(t_k) z_d(t_k)^T \right)^{-1},$$

provided of course that $\sum_{k=1}^l n_k z_d(t_k) z_d(t_k)^T \succ 0$. In the rest of this section, we consider the case where we would like to minimize the 2-norm of the matrix $\Sigma(\xi)$, which is the same as minimizing its largest eigenvalue. To avoid having to deal with the inverse that appears in its expression however, we will instead maximize the smallest eigenvalue of its inverse, or equivalently, if we define the following quantity,

$$F(\xi) := \sum_{k=1}^{l} w_k z_d(t_k) z_d(t_k)^T = \frac{\sigma^2}{m} \Sigma(\xi)^{-1},$$

maximize the minimum eigenvalue of $F(\xi)$. The matrix $F(\xi)$ is a well-known quantity in statistics called the *Fisher information matrix* of the design matrix ξ and maximizing its minimum eigenvalue corresponds to a common notion of optimality[2] in experimental design, that of *E-optimality.* Hence, the problem of interest becomes $\max_{\xi \text{ as in (7.1)}} \lambda_{\min} F(\xi)$.

7.2. Relaxing the optimal design problem using sum of squares. By dropping the condition that $n_k \in \mathbb{N}$, the optimal design problem can be rewritten as

$$\begin{aligned} &\max_{w_k, t_k, \gamma} \gamma \\ &\text{s.t.} \sum_{k=1}^{l} w_k z_d(t_k) z_d(t_k)^T \succeq \gamma I, w_k \geq 0, \sum_{k=1}^{l} w_k = 1. \end{aligned} \tag{7.3}$$

where I is the identity matrix. The matrix $\sum_{k=1}^{l} w_k z_d(t_k) z_d(t_k)^T$ is of size $|\mathbb{N}_d^n| \times |\mathbb{N}_d^n|$. We index it by (α, β) where $\alpha, \beta \in \mathbb{N}_d^n$. Note that entry (α, β) of the matrix is exactly

$$\int_{\mathcal{X}} x^\alpha x^\beta d\delta, \text{ where } \delta(x) = \sum_{k=1}^{l} w_k \mathbf{1}_{x=t_k}(x).$$

In other words, entry (α, β) of the matrix is the $\alpha + \beta$ moment of the Dirac measure δ at t_i. Define $y_\alpha := \int_{\mathcal{X}} x^\alpha d\delta, \alpha \in \mathbb{N}_{2d}^n$ and let $M(y)$ be the $\mathbb{N}_d^n \times \mathbb{N}_d^n$ matrix with entry (α, β) given by $y_{\alpha+\beta}$. It follows that (7.3) can be rewritten as:

$$\begin{aligned} &\max_{w_k, t_k, \gamma, y} \gamma \\ &\text{s.t. } M(y) \succeq \gamma I \\ &\qquad y_0 = 1, y_\alpha = \int_{\mathcal{X}} x^\alpha d\delta \text{ for } \alpha \in \mathbb{N}_{2d}^n. \end{aligned} \tag{7.4}$$

Similarly to (4.1), we now define the following set:

(7.5)

$$\mathcal{M}_{n,2d}(\mathcal{X}) := \{\{y_\alpha\}_{\alpha \in \mathbb{N}_{2d}^n} \mid \exists \text{ a measure } \mu \text{ on } \mathcal{X} \text{ such that } y_\alpha = \int_{\mathcal{X}} x^\alpha d\mu, \ \forall \alpha \in \mathbb{N}_{2d}^n\}.$$

Note that contrarily to (4.1), we are considering measures over $\mathcal{X}$ and not over $\mathbb{R}^n$. Hence, the dual of this set is the set of polynomials p with coefficients $\{p_\alpha\}_{\alpha \in \mathbb{N}_{2d}^n}$ such that p is nonnegative *over* $\mathcal{X}$, and not over $\mathbb{R}^n$ as previously; see [**61**, Section 4.4] for a proof. Using this definition and replacing the specific Dirac measure we

[2]There are many different ways of defining optimality, based essentially on minimizing various norms of $\Sigma(\xi)$; we refer the interested reader to [**31**] for more information on this topic.

were considering by a general measure μ over $\mathcal{X}$, we are able to further relax (7.4) to a problem that only depends on γ and y:

$$\begin{aligned} &\max_{\gamma,y} \gamma \\ &\text{s.t. } M(y) \succeq \gamma I \\ &\quad\; y_0 = 1, y \in \mathcal{M}_{n,2d}(\mathcal{X}). \end{aligned} \tag{7.6}$$

PROPOSITION 7.1. *Consider the optimization problem*

$$\begin{aligned} &\min_{\lambda\in\mathbb{R}, Q\in\mathbb{N}_d^n\times\mathbb{N}_d^n} \lambda \\ &s.t. \; \lambda - z_d(x)^T Q z_d(x) \geq 0, \forall x \in \mathcal{X} \\ &\quad\; tr(Q) = 1, Q \succeq 0, \end{aligned} \tag{7.7}$$

where $z_d(x) = (1, x_1, \ldots, x_n, \ldots, x_n^d)$ *is the standard vector of monomials. Then weak duality holds between* (7.6) *and* (7.7).

Strong duality holds under certain mild conditions, see [**61**, Section 4]. As is, (7.7) is not a tractable problem to solve. However, if one replaces the condition that $\lambda - z_d(x)^T Q z_d(x)$ be nonnegative over $\mathcal{X}$ by certificates of nonnegativity of the polynomial over $\mathcal{X}$ involving sum of squares polynomials, then the problem becomes a semidefinite program. One can proceed similarly in the primal (7.6) by relying on outer-approximations to the set $\mathcal{M}_{n,2d}(\mathcal{X})$. From the optimal solution y^* to the relaxation of the primal, one can recover—under some assumptions—a measure μ over $\mathcal{X}$ whose moments correspond to y^*. As it turns out, this measure is *atomic* which means that it can be written as a measure over a finite set of points (or *atoms*), each point being associated to a certain weight. We then pick our points t_k to be these atoms, and the corresponding weights to be the atom weights [**31**].

Part 4. Other applications

In this last part, we briefly touch upon some other important applications of sum of squares polynomials in optimization and game theory. We wish to stress that, though we have covered a large range of applications in this paper, we have by no means covered all of them. Other important applications of sum of squares techniques which are not included here are for example automated theorem proving [**47, 79**], extremal combinatorics [**85**], filer design [**41, 86**], and quantum information theory[**28, 33**]. We refer the reader to the aforementioned references if these topics are of interest.

8. Optimization

As has already been seen in this volume, as well as briefly in Section 4.3, sum of squares polynomials are widely used to tackle polynomial optimization problems (POPs), i.e., optimization problems where the objective is a polynomial and the constraints are given by polynomial equalities and inequalities. Though a major application of sum of squares techniques, we won't dwell on POPs as the topic has already been covered. We present however two other areas of optimization, namely copositive programming and robust semidefinite programming, where sum of squares techniques come into play. Multistage optimization is also an area of application of these techniques though not covered here; see [**15**] for more information.

8.1. Copositive programming.

DEFINITION 8.1. An $n \times n$ matrix M is said to be *copositive* if $x^T Mx \geq 0$, for all $x \in \mathbb{R}_+^n$.

Without loss of generality, it can be assumed that M is symmetric. We denote by C_n the set of $n \times n$ copositive matrices. This is a proper cone in the space of symmetric $n \times n$ matrices. At first glance, this definition may seem similar to the definition of positive semidefiniteness. A major difference between the two is that checking membership of a matrix to the set of positive semidefinite matrices can be done in polynomial time; checking membership of a matrix to C_n however is NP-hard [**71**]. We refer the reader to [**36**] and the references therein for more information on copositive programming, including the dual cone to C_n and applications of copositive programming. We will cover one example here relating to the stability number of a graph.

8.1.1. *Copositive matrices and sum of squares polynomials.* As it is NP-hard to check membership to C_n, it is of interest to develop sufficient conditions for membership to C_n that are checkable in polynomial time. We give an example of such a condition next.

PROPOSITION 8.2. *Let M be an $n \times n$ symmetric matrix. If $M = P + N$ for some matrices $P \succeq 0$ and $N \geq 0$, then M is copositive.*

Clearly checking whether M satisfies this condition is a semidefinite (feasibility) program. How strong is this sufficient condition? It turns out that for $n < 5$, the set of matrices $\{M \mid M = P + N, P \succeq 0, N \geq 0\}$ exactly coincides with C_n. For $n \geq 5$, this is not true anymore, as evidenced by the so-called Horn matrix, a 5×5 zero-one matrix which is copositive but cannot be written as $P + N$, $P \succeq 0$, $N \geq 0$; see, e.g., [**36**]. Better than one sufficient condition, however, would be a hierarchy of sufficient conditions, with each level giving rise to an improved inner approximation of C_n. This can be obtained by relating the notion of copositivity back to polynomial nonnegativity and then using sum of squares-based approximations. How can we do this in practice? First, it is straightforward to see that M is copositive if and only if the associated quartic form

$$m(x) := \sum_{i,j} m_{ij} x_i^2 x_j^2 \tag{8.1}$$

is globally nonnegative. This leads to a natural sufficient condition for copositivity: requiring that $m(x)$ in 8.1 be a sum of squares rather than nonnegative. Interestingly enough, this first sufficient condition is equivalent to that given in Proposition 8.2.

PROPOSITION 8.3. [**76**, Section 5] *A matrix M can be written as $P + N$, with $P \succeq 0$ and $N \geq 0$ if and only if the associated quartic form $m(x)$ is a sum of squares.*

For the proof, we refer the reader to [**76**]. Furthermore, this particular rewriting of our initial sufficient condition clues us in on how to construct an improving hierarchy of semidefinite-based inner approximations to the set of copositive matrices; see [**76**]. Indeed, let

$$K_r := \{M \in \mathbb{R}^{n \times n} \mid \left(\sum_{i=1}^n x_i^2\right)^r \cdot \left(\sum_{i,j}^n m_{ij} x_i^2 x_j^2\right) \text{ is sos}\}.$$

It is easy to see that K_0 is the set $\{M \in \mathbb{R}^{n\times n} \mid M = P + N, P \succeq 0, N \geq 0\}$, that $K_r \subseteq K_{r+1}, \forall r \geq 0$, and that $K_r \subseteq C_n$, $\forall r \geq 0$. Testing membership to K_r is a semidefinite feasibility program, whose size grows with r. An important property of these inner approximations is that they get arbitrarily close to C_n, i.e., $int(C_n) \subseteq \cup_{r\in\mathbb{N}} K_r$. The latter fact is a consequence of a theorem by Polyá [**81**].

8.1.2. *Using copositive programming to approximate the stability number of the graph.* Let $G = (V, E)$ be a graph on n nodes. A stable set of the graph G is a subset of its nodes, no two of which have an edge between them. The stability number $\alpha(G)$ of G is the size of the largest stable set in G. The ability to compute $\alpha(G)$ has applications in scheduling and coding theory among other areas. The issue however is that $\alpha(G)$ is hard to compute, though reasonable upper bounds on it can often be obtained using linear programming or semidefinite programming. One particularly well-known semidefinite programming approximation is due to Lovász [**65**]; the upper bound obtained via this method is denoted by $\vartheta(G)$ and called the *theta number*. It can be shown that $\vartheta(G)$ is always an improvement on the bounds obtained via the standard linear programming relaxation. In [**32**], the authors propose a copositive-programming formulation of the stability number, the relaxation of which can beat $\vartheta(G)$.

PROPOSITION 8.4. [**32**, Corollary 2.4] *For any graph $G = (V, E)$ with adjacency matrix A, one has*

$$\begin{aligned}\alpha(G) = &\min_{\lambda} \lambda \\ &s.t.\ \lambda(I + A) - ee^T \in C_n,\end{aligned}$$

where e is the $n \times 1$ vector of ones.

Though this reformulation does not make the problem any more tractable, it does open the door to using the aforementioned approximations of the copositive cone, optimization over which is tractable. This gives rise to a hierarchy of optimization problems indexed by r:

$$\begin{aligned}\vartheta_r'(G) := &\min_{\lambda} \lambda \\ &\text{s.t. } \lambda(I + A) - ee^T \in K_r.\end{aligned}$$

Already at level $r = 0$, the authors are able to show that $\vartheta_0'(G)$ improves on $\vartheta(G)$. Furthermore, it is shown that if $r \geq \alpha^2(G)$, then $\vartheta_r'(G)$ is equal to the true stability number. We refer the reader to [**32**] for more details.

8.2. Robust semidefinite programming. We consider the following optimization problem:

$$\begin{aligned}u_{opt} := &\min_{y\in\mathbb{R}^n} c^T y \\ &\text{s.t. } F(x, y) \succ 0, \forall x \in \mathbb{R}^m \text{ with } G(x) \preceq 0,\end{aligned} \tag{8.2}$$

where c is an $n \times 1$ vector, F is a $p \times p$ symmetric matrix whose entries depend polynomially on x and affinely on y, and $G(x)$ is a $q \times q$ symmetric matrix whose entries depend polynomially on x. We say that F and G are *polynomial matrices.* Note that for fixed x, (8.2) is a semidefinite program. However, unlike other semidefinite programs, we require here that y satisfy $F(x, y) \succ 0$, regardless of the value that x takes within the set $\{x \in \mathbb{R}^n \mid G(x) \preceq 0\}$. We can view x as encoding uncertainty in our problem and we want our solution y to be robust to this uncertainty. Scenarios where one would encounter semidefinite programs with uncertainty are numerous; see [**14**, **37**, **89**] for some examples. We follow the approach described in [**89**] for this problem. It relies on the concept of *sum of squares matrix* that we define next.

DEFINITION 8.5. A $p \times p$ matrix $S(x)$ whose entries are polynomials in $x \in \mathbb{R}^m$ is said to be a *sum of squares (sos) matrix* if there exists a $n \times p$ polynomial matrix $T(x)$ such that

$$S(x) = T(x)^T T(x).$$

An equivalent definition is for the scalar-valued polynomial $y^T S(x) y$ be a sum of squares polynomial in the variables $(x, y) \in \mathbb{R}^m \times \mathbb{R}^p$. As a consequence, constraining a matrix to be a sum of squares matrix can be imposed using semidefinite programming.

REMARK 8.6. This may seem similar to the concept of sos-convexity, which we saw in Section 6. As it turns out, a polynomial is sos-convex if and only if its Hessian is an sos matrix.

The difficulty in solving (8.2) lies in the "for all x" quantifier. A key step is then to reformulate the constraint in such a way that the quantifier does not appear anymore. This is done by making use of a bilinear mapping $(\cdot, \cdot)_p : \mathbb{R}^{pq \times pq} \times \mathbb{R}^{q \times q} \mapsto \mathbb{R}^{p \times p}$, such that $(A, B)_p = tr_p(A^T(I_p \otimes B))$, where $\otimes$ is the Kronecker product and

$$tr_p(C) = \begin{bmatrix} tr(C_{11}) & \dots & tr(C_{1p}) \\ \vdots & \ddots & \vdots \\ tr(C_{p1}) & \dots & tr(C_{pp}) \end{bmatrix} \quad \text{with } C \in \mathbb{R}^{pq \times pq}, C_{jk} \in \mathbb{R}^{q \times q}.$$

PROPOSITION 8.7. *Let $A \in \mathbb{R}^{pq \times pq}$ and $B \in \mathbb{R}^{q \times q}$. If $A \succeq 0$ and $B \succeq 0$, then $(A, B)_p \succeq 0$.*

Using this bilinear mapping, we define a new optimization problem:

$$\begin{aligned} v_{opt} := \min_{y, \epsilon, S} \quad & c^T y \\ \text{s.t.} \quad & \epsilon > 0, \\ & F(x, y) + (S(x), G(x))_p - \epsilon I_p \text{ is an sos matrix,} \\ & S(x) \text{ is an sos matrix.} \end{aligned} \tag{8.3}$$

where $S(x)$ is a $pq \times pq$ matrix with polynomial entries. which is related to our robust semidefinite programming problem in (8.2). This is made explicit in the Proposition below.

PROPOSITION 8.8. [**89**] *Let u_{opt} and v_{opt} be defined as in* (8.2) *and* (8.3). *We have $u_{opt} \leq v_{opt}$.*

PROOF. Suppose y, ϵ, and $S(x)$ are feasible for (8.3) and let x_0 be an arbitrary point such that $G(x_0) \preceq 0$. Using the third constraint of (8.3), we have that $S(x_0) \succeq 0$. From the second constraint in (8.3) and Proposition 8.7, it follows that

$$F(x_0, y) - \epsilon I_p \succeq -(S(x_0), G(x_0))_p = (S(x_0), -G(x_0))_p \succeq 0$$

and so $F(x_0, y) \succ 0$. This implies that y is feasible for (8.3) as x_0 was chosen arbitrarily, and so that $u_{opt} \leq v_{opt}$. □

Under a modest technical condition on $G(x)$ (a generalization of the so-called Archimedean condition to matrices), it can in fact be shown that $u_{opt} = v_{opt}$; see [**89**, Theorem 1]. Though (8.3) is not tractable as is, it can easily be made tractable by fixing the degree of the polynomials that appear in the entries of $S(x)$. Indeed, it then becomes a semidefinite program. Under the previous assumption, and as the degree of the polynomials in S grows, the optimal values of the approximations converge to u_{opt}. Once again, we refer the reader to [**89**] for more details.

9. Game theory: polynomial games

9.1. Polynomial games and Nash equilibria. In a game, multiple agents (the *players*) have to make decisions, each guided by his or her preferences, and each influenced by the decisions of the other players. Game theory is the mathematical theory that studies what behavior emerges in these situations; see, e.g., [**73**]. We consider *strategic form games* which are games specified by three components: the number of players, the set of actions (or pure strategies) for each player, and their payoffs. In our set-up, we have two players, Player 1 and Player 2, each having a set of actions S_1 and S_2, and a payoff function $P_i : S_1 \times S_2 \to \mathbb{R}$, $i = 1, 2$. The sets S_i, $i = 1, 2$ can generally be finite or infinite; in our case we will take $S_1 = [-1, 1]$ and $S_2 = [-1, 1]$. Furthermore, we assume that the payoff function of Player 1 is a polynomial in actions $x \in S_1$ and $y \in S_2$:

$$P_1(x, y) = \sum_{i=0}^{n} \sum_{j=0}^{m} p_{ij} x^i y^j,$$

and that the payoff function of Player 2 is a polynomial in actions $x \in S_1$ and $y \in S_2$. This gives rise to the name *polynomial game.*

In strategic games, when player i plays, (s)he can decide to play one action in a deterministic fashion—a *pure* strategy—or (s)he can decide to pick a strategy that doesn't assign all mass to one strategy, but rather a probability distribution over the set of strategies S_i. The latter is called a *mixed* strategy. We assume that each player randomizes their strategy independently of the other. Our goal is to efficiently compute a pair of mixed strategies, also called a *strategy profile*, that is a Nash equilibrium. We use $\Lambda(S)$ to denote the set of probability measures over S and by $\mu \times \nu$ the product measure of two measures μ and ν.

DEFINITION 9.1. Consider a two-player polynomial game with payoff functions $P_1(x, y), P_2(x, y)$. Let μ^* (resp. ν^*) be a probability measure over S_1 (resp. S_2). A mixed strategy profile (μ^*, ν^*) is a Nash equilibrium if $E_{\mu^* \times \nu^*}[P_1(x, y)] \geq E_{\mu \times \nu^*}[P_1(x, y)], \forall \mu \in \Lambda(S_1)$ and $E_{\mu^* \times \nu^*}[P_2(x, y)] \geq E_{\mu^* \times \nu}[P_2(x, y)], \forall \nu \in \Lambda(S_2)$.

In layman's terms, this means that if one player changes his strategy while the strategy of the other player remains the same, his or her expected payoff cannot increase. Thus, there is no incentive for any one player to unilaterally change his or her strategy.

REMARK 9.2. In general, Nash equilibria do not necessarily exist, though there are a number of games for which existance is guaranteed. Nash himself showed that there always exists a Nash equilibrium for games that involve a finite number of players and a finite number of pure strategies. Existence of these equilibria does not guarantee however that they are easy to compute: in fact, it is believed that computing Nash equilibria is a hard problem, even in the simple 2-player finite-strategy setting; see [**27**, **30**]. There are a few specific cases where Nash equilbria computation can be done in polynomial time—when a 2-player finite-strategy game is zero-sum for example [**29**]. In our case, we will see that not only do Nash equilibria exist, but they can be computed using semidefinite programming.

We further assume that our game is zero sum, i.e., $P_1 = -P_2$. For ease of notation, we consequently drop the subscript and write P instead of P_1 (and $-P$ instead of P_2). This is the simplest case of a polynomial game, first introduced in [**34**]. From the zero-sum game assumption, the following equivalence is immediate.

PROPOSITION 9.3. *Consider a zero-sum polynomial game with payoff function $P(x, y)$. Let μ^* (resp. ν^*) be a probability measure over S_1 (resp. S_2) The strategy profile (μ^*, ν^*) is a Nash equilibrium if and only if (μ^*, ν^*) satisfies the saddle-point condition*

(9.1)

$$E_{\mu\times\nu^*}[P(x,y)] \leq E_{\mu^*\times\nu^*}[P(x,y)] \leq E_{\mu^*\times\nu}[P(x,y)], \ \forall \mu \in \Lambda(S_1), \forall \nu \in \Lambda(S_2).$$

9.2. Computing Nash equilibria using semidefinite programming. We denote by $\mathcal{M}_1([-1,1])$ the set of moments which have a representing measure over $[-1,1]$. This is similar to the notation employed in Section 4 for the global setting. The main theorems in [**77**], from which this section is inspired, state the following.

THEOREM 9.4. [**77**, Theorem 2.2-Theorem 2.7] *Consider a zero-sum polynomial game with payoff function $P(x, y)$. There exist mixed strategies (μ^*, ν^*) verifying the saddle-point condition* (9.1). *Furthermore, these strategies can be obtained by solving the primal-dual pair of optimization problems:*

$$\begin{aligned} \max_{\mu_0,\dots,\mu_n,\lambda} \quad & \lambda \\ s.t. \quad & \sum_{j=0}^{m}\Big(\sum_{i=0}^{n} p_{ij}\mu_i\Big)y^j - \lambda \geq 0, \forall y \in [-1,1], \\ & (\mu_0,\dots,\mu_n) \in \mathcal{M}_1([-1,1]),\ \mu_0 = 1 \end{aligned} \tag{9.2}$$

$$\begin{aligned} \min_{\nu_0,\dots,\nu_m,\gamma} \quad & \gamma \\ s.t. \quad & \gamma - \sum_{i=0}^{n}\Big(\sum_{j=0}^{m} p_{ij}\nu_j\Big)x^i \geq 0, \forall x \in [-1,1] \\ & (\nu_0,\dots,\nu_m) \in \mathcal{M}_1([-1,1]),\ \nu_0 = 1. \end{aligned} \tag{9.3}$$

In the previous optimization problems, $(\mu_0, \dots, \mu_n)$ and $(\nu_0, \dots, \nu_m)$ are moments of a measure over $[-1,1]$. As the measures considered are over intervals and the polynomials are univariate, the constraints of these problems have a tractable semidefinite representation. We refer the reader to [**77**] for the exact semidefinite programming formulation. From the moments $(\mu_0, \dots, \mu_n)$ and $(\nu_0, \dots, \nu_m)$ obtained, one can extract corresponding *atomic* measures μ^*, ν^* over $[-1,1]$, which are measures over a finite number of points. These measures constitute our desired strategies. Again, the extraction process is not covered here but can be found, e.g., in [**19**, Section 3.3.5].

This theorem gives us both the existence of a Nash equilibrium and the means to compute it efficiently. Before we proceed, we give some intuition as to how these optimization problems arise. It is straightforward to show that for the game at hand, (μ^*, ν^*) satisfies the saddle-point condition (9.1) if and only if

$$\max_{\mu\in\Lambda(S_1)} \min_{\nu\in\Lambda(S_2)} E_{\mu\times\nu}[P(x,y)] = \min_{\nu\in\Lambda(S_2)} \max_{\mu\in\Lambda(S_1)} E_{\mu\times\nu}[P(x,y)]. \tag{9.4}$$

and

$$\mu^* \in \arg \max_{\mu\in\Lambda(S_1)} \min_{\nu\in\Lambda(S_2)} E_{\mu\times\nu}[P(x,y)], \quad \nu^* \in \arg \min_{\nu\in\Lambda(S_2)} \max_{\mu\in\Lambda(S_1)} E_{\mu\times\nu}[P(x,y)].$$

We call the common optimal value the *value* of the game, μ^* a *maximin* strategy and ν^* a *minimax* strategy. These strategies are intuitive: as $P(x,y)$ is the payment from Player 2 to Player 1, it follows that Player 1 would try to maximize the payoff in the worst case scenario that Player 2 has played a strategy that minimizes it as much as possible, and conversely. How does this equivalent characterization help us obtain the optimization problems in (9.2) and (9.3)? It suffices to simply rewrite the problems that appear in (9.4). Consider, for example,

$$\max_{\mu\in\Lambda(S_1)} \min_{\nu\in\Lambda(S_2)} E_{\mu\times\nu}[P(x,y)]. \tag{9.5}$$

We use the fact that $P(x,y)$ is a polynomial in x and y to express the expectation as a polynomial in its moments: we obtain $E_{\mu\times\nu}[P(x,y)] = \sum_{i=0}^n \sum_{j=0}^m p_{ij}\mu_i\nu_j$ where mu_i (resp. ν_j) is the ith (resp. j^{th}) moment of μ (resp. ν). As a consequence, (9.5) can be written as:

$$\begin{array}{l} \max\limits_{(\mu_0,\dots,\mu_n)\in\mathcal{M}_{1,n}([-1,1])} \begin{bmatrix} \min_{(\nu_0,\dots,\nu_m)\in\mathcal{M}_{1,m}([-1,1])} \sum_{i=0}^n \sum_{j=0}^m p_{ij}\mu_i\nu_j \\ \text{s.t. } \nu_0 = 1 \end{bmatrix} \\ \text{s.t. } \mu_0 = 1, \end{array}$$

where $\mathcal{M}_{1,n}([-1,1])$ is once again defined as the set of (truncated) moment sequences that have a representative measure over $[-1,1]$. We now use duality to rewrite the inner problem: note that this problem is very similar to (4.4) and so its dual will be of a similar form to (4.3), the only difference being that the measures are over $[-1,1]$. Problem (9.5) then becomes:

$$\begin{aligned} \max_{\mu_i,\lambda} \quad & \lambda \\ \text{s.t.} \quad & \sum_{j=0}^m (\sum_{i=0}^n p_{ij}\mu_i) y^j - \lambda \geq 0, \forall y \in [-1,1] \\ & (\mu_0,\dots,\mu_n) \in \mathcal{M}_{1,n}([-1,1]), \ \mu_0 = 1, \end{aligned}$$

which corresponds to (9.2). One can proceed likewise for (9.3).

Theorem 9.4 is a nice generalization of two-player zero-sum finite games where computing a Nash equilibrium amounts to solving a pair of primal-dual linear programs. An extension of polynomial games to the multivariate setting can be found in [**57**]; other extensions to, e.g., different notions of equilibria such as correlated equilibria [**93**] or to polynomial stochastic games [**91**] can be found in the literature.

Part 5. Conclusion: a word on implementation challenges

A topic that is central to applications but that we have barely touched upon so far is how to implement these methods in practice. In particular, what software should we use to solve sum of squares programs? There are two components here to consider: which semidefinite programming solver to use and which parser to use which converts the sum of squares program to a semidefinite program. Indeed, in theory, one could manually convert the sum of squares program at hand into a semidefinite program, but this can be quite tedious. It is consequently much more convenient to use a parser whose role is to automate this process. Note however that not all parsers interface with all solvers, and that one need sometimes access parsers or solvers within other software or interfaces (e.g., MATLAB).

We start by reviewing a few semidefinite programming solvers (the list is by no means meant to be exhaustive). Choosing a good-quality solver is a crucial step in coming up with robust solutions to the problem at hand in a reasonable amount of time. MOSEK [**1**], SDPT3 [**95**], Sedumi [**94**], and SDPA [**100**] are established solvers, with the first being a commercial solver (free with an academic license), and the latter three being free. SDPA interfaces with Python, MOSEK interfaces with C, Java, Python, and MATLAB, and Sedumi and SDPT3 interface with MATLAB. All of these solvers rely on interior point methods, which unfortunately do not always scale very well with the size of the problem. In light of this, new solvers have been developed which rely on augmented Lagrangian methods instead, such as SDPNAL/SDPNAL+ [**101**], CDCS [**102**], and SCS [**72**]. These are all free. The first uses Newton-Conjugate Gradient augmented Lagrangian, whereas the latter two use ADMM. Furthermore SDPNAL/SDPNAL+ and CDCS can be accessed from MATLAB, whereas SCS is written in C and can be used in other C, C++, Python environments, as well as in MATLAB, R and Julia.

In terms of parsers, the most commonly used are perhaps YALMIP [**64**], Julia [**18**], SOSTOOLS [**83**], SPOT [**70**], Gloptipoly [**50**], and Macaulay2 [**43**]. YALMIP is a toolbox for modelling and optimization in MATLAB that has a special sos module. It can be interfaced with many solvers including MOSEK, SDPT3 and Sedumi. Julia is an open-source programming language for technical computing. Optimization is done via its modeling package JuMP [**35**] and sum of squares problems can be tackled via one of its packages SumofSquares.jl [**26**]. SOSTOOLS is a free MATLAB toolbox for sos programs. It can be interfaced with a number of solvers including SeDuMi, SDPT3, CSDP, SDPNAL, SDPNAL+, CDCS and SDPA. SPOT can be viewed as an alternative to SOSTOOLS as it is also a MATLAB toolbox for sos programs, but its focus is mainly towards control theory. GloptiPoly is a free MATLAB toolbox, which solves what is known as the *generalized moment problem*. This includes the moment problem as described in Section 4, and hence can be used to tackle polynomial optimization problems. It can be interfaced with many different solvers including Sedumi, SDPT3 and MOSEK. Finally, Macaulay2 is a free computer algebra system geared towards research in algebraic geometry. However, via a package, it can be used to solve sum of squares programs.

As mentioned above, a direction currently taken in solver development involves replacing interior point methods by methods that can robustly solve very large problems, such as ADMM [**25**]. This is due to the fact that the semidefinite programs generated by sum of squares programs are very large, sometimes prohibitively large for the applications considered. This is arguably the largest bottleneck for the dissemination of sum of squares techniques, and a large swath of research now focuses on making sos programs more scalable. This can be done by constructing new solvers for semidefinite programs that rely on more scalable algorithms as we saw above. Another research direction, complementary to this one, is to leverage the structure of the semidefinite program at hand to reduce its size. Structures of interest can include e.g. symmetries in the problem or sparsity; see [**40, 96, 97**] for some of these directions. A very different research direction involves replacing the semidefinite program at hand by cheaper conic programs with trade-offs in accuracy; see [**8, 98**] for some examples of this direction. The hope is that by combining these different research directions, one will be able to tackle large-scale sos programs and open up many new areas to the use of sos programming.

Acknowledgments

The author would like to thank Amir Ali Ahmadi for painstakingly going over an initial version of this manuscript to provide the author with feedback. His very relevant comments considerably improved this draft.

References

[1] *MOSEK reference manual*, 2013. Version 7. Latest version available at `http://www.mosek.com/`.

[2] A. A. Ahmadi, *Non-monotonic Lyapunov functions for stability of nonlinear and switched systems: theory and computation*, Master's thesis, Massachusetts Institute of Technology, June 2008. Available at `http://aaa.lids.mit.edu/publications`.

[3] A. A. Ahmadi, M. Curmei, and G. Hall, *Shape-constrained regression and nonnegative polynomials*, in preparation, 2019.

[4] A. A. Ahmadi and G. Hall. *On the complexity of detecting convexity over a box*, to appear in Mathematical Programming, Series A, 2018. Available at `arXiv:1806.06173`.

[5] A. A. Ahmadi, G. Hall, A. Makadia, and V. Sindhwani, *Geometry of 3D environments and sum of squares polynomials*, Proceedings of Robotics: Science and Systems, 2017. Available at `arXiv:1611.07369`.

[6] Amir Ali Ahmadi, Raphaël M. Jungers, Pablo A. Parrilo, and Mardavij Roozbehani, *Joint spectral radius and path-complete graph Lyapunov functions*, SIAM J. Control Optim. **52** (2014), no. 1, 687–717, DOI 10.1137/110855272. MR3168608

[7] A. A. Ahmadi, M. Krstic, and P. A. Parrilo, *A globally asymptotically stable polynomial vector field with no polynomial Lyapunov function*, Proceedings of the 50th IEEE Conference on Decision and Control, 2011.

[8] Amir Ali Ahmadi and Anirudha Majumdar, *DSOS and SDSOS optimization: more tractable alternatives to sum of squares and semidefinite optimization*, SIAM J. Appl. Algebra Geom. **3** (2019), no. 2, 193–230, DOI 10.1137/18M118935X. MR3939321

[9] A. A. Ahmadi and P. A. Parrilo, *Converse results on existence of sum of squares Lyapunov functions*, Proceedings of the 50th IEEE Conference on Decision and Control, 2011.

[10] Amir Ali Ahmadi and Pablo A. Parrilo, *A complete characterization of the gap between convexity and sos-convexity*, SIAM J. Optim. **23** (2013), no. 2, 811–833, DOI 10.1137/110856010. MR3049657

[11] Panos J. Antsaklis and Anthony N. Michel, *Linear systems*, Birkhäuser Boston, Inc., Boston, MA, 2006. MR2197279

[12] Andrea Bacciotti and Lionel Rosier, *Liapunov functions and stability in control theory*, 2nd ed., Communications and Control Engineering Series, Springer-Verlag, Berlin, 2005. MR2146587

[13] Boaz Barak and David Steurer, *Sum-of-squares proofs and the quest toward optimal algorithms*, Proceedings of the International Congress of Mathematicians—Seoul 2014. Vol. IV, Kyung Moon Sa, Seoul, 2014, pp. 509–533. MR3727623

[14] A. Ben-Tal and A. Nemirovski, *Robust truss topology design via semidefinite programming*, SIAM J. Optim. **7** (1997), no. 4, 991–1016, DOI 10.1137/S1052623495291951. MR1479611

[15] Dimitris Bertsimas, Dan Andrei Iancu, and Pablo A. Parrilo, *A hierarchy of near-optimal policies for multistage adaptive optimization*, IEEE Trans. Automat. Control **56** (2011), no. 12, 2809–2824, DOI 10.1109/TAC.2011.2162878. MR2906670

[16] Dimitris Bertsimas and Ioana Popescu, *On the relation between option and stock prices: a convex optimization approach*, Oper. Res. **50** (2002), no. 2, 358–374, DOI 10.1287/opre.50.2.358.424. MR1899463

[17] Dimitris Bertsimas and Ioana Popescu, *Optimal inequalities in probability theory: a convex optimization approach*, SIAM J. Optim. **15** (2005), no. 3, 780–804, DOI 10.1137/S1052623401399903. MR2142860

[18] Jeff Bezanson, Alan Edelman, Stefan Karpinski, and Viral B. Shah, *Julia: a fresh approach to numerical computing*, SIAM Rev. **59** (2017), no. 1, 65–98, DOI 10.1137/141000671. MR3605826

[19] G. Blekherman, P. A. Parrilo, and R. Thomas, *Semidefinite Optimization and Convex Algebraic Geometry*, SIAM Series on Optimization, 2013.

[20] Vincent D. Blondel, *The birth of the joint spectral radius: an interview with Gilbert Strang*, Linear Algebra Appl. **428** (2008), no. 10, 2261–2264, DOI 10.1016/j.laa.2007.12.010. MR2405243

[21] V. D. Blondel, J. M. Hendrickx, A. Olshevsky, and J. N. Tsitsiklis, *Convergence in multiagent coordination, consensus, and flocking*, 44th IEEE Conference on Decision and Control, 2005, pages 2996–3000. IEEE.

[22] Vincent D. Blondel and Yurii Nesterov, *Polynomial-time computation of the joint spectral radius for some sets of nonnegative matrices*, SIAM J. Matrix Anal. Appl. **31** (2009), no. 3, 865–876, DOI 10.1137/080723764. MR2538656

[23] Vincent D. Blondel and John N. Tsitsiklis, *A survey of computational complexity results in systems and control*, Automatica J. IFAC **36** (2000), no. 9, 1249–1274, DOI 10.1016/S0005-1098(00)00050-9. MR1834719

[24] Stephen Boyd, Laurent El Ghaoui, Eric Feron, and Venkataramanan Balakrishnan, *Linear matrix inequalities in system and control theory*, SIAM Studies in Applied Mathematics, vol. 15, Society for Industrial and Applied Mathematics (SIAM), Philadelphia, PA, 1994. MR1284712

[25] S. Boyd, N. Parikh, E. Chu, B. Peleato, and J. Eckstein, *Distributed optimization and statistical learning via the alternating direction method of multipliers*, Foundations and Trends® in Machine learning, **3** no. 1, 2011, 1–122.

[26] R. D. Brackston, A. Wynn, and M. P. H. Stumpf, *Construction of quasipotentials for stochastic dynamical systems: an optimization approach*, Phys. Rev. E **98** (2018), no. 2, 022136, 12, DOI 10.1103/physreve.98.022136. MR3862388

[27] X. Chen and X. Deng, *Settling the complexity of two-player Nash equilibrium*, 2006 47th Annual IEEE Symposium on Foundations of Computer Science (FOCS'06), 2006, pages 261–272. IEEE.

[28] A. M. Childs, A. J. Landahl, and P. A. Parrilo, *Quantum algorithms for the ordered search problem via semidefinite programming*, Physical Review A, **75** no. 3, 032335, 2007.

[29] George B. Dantzig, *A proof of the equivalence of the programming problem and the game problem*, Activity Analysis of Production and Allocation, Cowles Commission Monograph No. 13, John Wiley & Sons, Inc., New York, N. Y.; Chapman & Hall, Ltd., London, 1951, pp. 330–335. MR0046019

[30] Constantinos Daskalakis, Aranyak Mehta, and Christos Papadimitriou, *A note on approximate Nash equilibria*, Theoret. Comput. Sci. **410** (2009), no. 17, 1581–1588, DOI 10.1016/j.tcs.2008.12.031. MR2510264

[31] Yohann De Castro, Fabrice Gamboa, Didier Henrion, Roxana Hess, and Jean-Bernard Lasserre, *Approximate optimal designs for multivariate polynomial regression*, Ann. Statist. **47** (2019), no. 1, 127–155, DOI 10.1214/18-AOS1683. MR3909929

[32] E. de Klerk and D. V. Pasechnik, *Approximation of the stability number of a graph via copositive programming*, SIAM J. Optim. **12** (2002), no. 4, 875–892, DOI 10.1137/S1052623401383248. MR1922500

[33] A. C. Doherty, P. A. Parrilo, and F. M. Spedalieri, *Complete family of separability criteria*, Physical Review A, **69** no. 2, 022308, 2004.

[34] M. Dresher, S. Karlin, and L. S. Shapley, *Polynomial games*, Contributions to the Theory of Games, Annals of Mathematics Studies, no. 24, Princeton University Press, Princeton, N. J., 1950, pp. 161–180. MR0039225

[35] Iain Dunning, Joey Huchette, and Miles Lubin, *JuMP: a modeling language for mathematical optimization*, SIAM Rev. **59** (2017), no. 2, 295–320, DOI 10.1137/15M1020575. MR3646493

[36] M. Dür. *Copositive programming–a survey*, Recent Advances in Optimization and its Applications in Engineering, Springer, 2010, pages 3–20.

[37] Laurent El Ghaoui, Francois Oustry, and Hervé Lebret, *Robust solutions to uncertain semidefinite programs*, SIAM J. Optim. **9** (1999), no. 1, 33–52, DOI 10.1137/S1052623496305717. MR1660106

[38] Gerald B. Folland, *How to integrate a polynomial over a sphere*, Amer. Math. Monthly **108** (2001), no. 5, 446–448, DOI 10.2307/2695802. MR1837866

[39] Shmuel Friedland and Lek-Heng Lim, *The computational complexity of duality*, SIAM J. Optim. **26** (2016), no. 4, 2378–2393, DOI 10.1137/16M105887X. MR3566920

[40] Karin Gatermann and Pablo A. Parrilo, *Symmetry groups, semidefinite programs, and sums of squares*, J. Pure Appl. Algebra **192** (2004), no. 1-3, 95–128, DOI 10.1016/j.jpaa.2003.12.011. MR2067190

[41] Y. Genin, Y. Hachez, Y. Nesterov, and P. Van Dooren, *Convex optimization over positive polynomials and filter design*, Proceedings of the 2000 UKACC International Conference on Control. Cambridge University, 2000.

[42] Paul J. Goulart and Sergei Chernyshenko, *Global stability analysis of fluid flows using sum-of-squares*, Phys. D **241** (2012), no. 6, 692–704, DOI 10.1016/j.physd.2011.12.008. MR2881107

[43] D. R. Grayson and M. E. Stillman, *Macaulay2: a software system for research in algebraic geometry.* Available at `http://www.math.uiuc.edu/Macaulay2/`.

[44] Maya Gupta, Andrew Cotter, Jan Pfeifer, Konstantin Voevodski, Kevin Canini, Alexander Mangylov, Wojciech Moczydlowski, and Alexander van Esbroeck, *Monotonic calibrated interpolated look-up tables*, J. Mach. Learn. Res. **17** (2016), Paper No. 109, 47. MR3543515

[45] G. Hall, *Optimization over Nonnegative and Convex Polynomials With and Without Semidefinite Programming*, PhD thesis, ORFE, Princeton University, 2018. Available at `arXiv:1806.06996`.

[46] Lauren A. Hannah and David B. Dunson, *Multivariate convex regression with adaptive partitioning*, J. Mach. Learn. Res. **14** (2013), 3261–3294. MR3144462

[47] John Harrison, *Verifying nonlinear real formulas via sums of squares*, Theorem proving in higher order logics, Lecture Notes in Comput. Sci., vol. 4732, Springer, Berlin, 2007, pp. 102–118, DOI 10.1007/978-3-540-74591-4_9. MR2550881

[48] D. Henrion and A. Garulli (eds.), *Positive polynomials in control*, Lecture Notes in Control and Information Sciences, vol. 312, Springer-Verlag, Berlin, 2005. MR2121436

[49] Didier Henrion and Jean-Bernard Lasserre, *Detecting global optimality and extracting solutions in GloptiPoly*, Positive polynomials in control, Lect. Notes Control Inf. Sci., vol. 312, Springer, Berlin, 2005, pp. 293–310, DOI 10.1007/10997703_15. MR2123528

[50] Didier Henrion, Jean-Bernard Lasserre, and Johan Löfberg, *GloptiPoly 3: moments, optimization and semidefinite programming*, Optim. Methods Softw. **24** (2009), no. 4-5, 761–779, DOI 10.1080/10556780802699201. MR2554910

[51] D. Huang, S. Chernyshenko, P. Goulart, D. Lasagna, O. Tutty, and F. Fuentes, *Sum-of-squares of polynomials approach to nonlinear stability of fluid flows: an example of application*, Proc. A. **471** (2015), no. 2183, 20150622, 18, DOI 10.1098/rspa.2015.0622. MR3436283

[52] Z. Jarvis-Wloszek, R. Feeley, W. Tan, K. Sun, and A. Packard, *Some controls applications of sum of squares programming*, Proceedings of the 42th IEEE Conference on Decision and Control, 2003, pages 4676–4681.

[53] Raphaël Jungers, *The joint spectral radius: Theory and applications*, Lecture Notes in Control and Information Sciences, vol. 385, Springer-Verlag, Berlin, 2009. MR2507938

[54] H. Khalil, *Nonlinear systems*, Prentice Hall, 2002, 3rd edition.

[55] M. Krstic, I. Kanellakopoulos, P. V. Kokotovic, et al, *Nonlinear and adaptive control design*, vol. 222. Wiley, New York, 1995.

[56] Y. Kurzweil, *On the inversion of the second theorem of Lyapunov on stability of motion*, Czechoslovak Math. J, **81**, no. 6, 1956, 217–259.

[57] R. Laraki and J. B. Lasserre, *Semidefinite programming for min-max problems and games*, Math. Program. **131** (2012), no. 1-2, Ser. A, 305–332, DOI 10.1007/s10107-010-0353-y. MR2886150

[58] Jean B. Lasserre, *Global optimization with polynomials and the problem of moments*, SIAM J. Optim. **11** (2000/01), no. 3, 796–817, DOI 10.1137/S1052623400366802. MR1814045

[59] Jean B. Lasserre, *Bounds on measures satisfying moment conditions*, Ann. Appl. Probab. **12** (2002), no. 3, 1114–1137, DOI 10.1214/aoap/1031863183. MR1925454

[60] Jean Bernard Lasserre, *An introduction to polynomial and semi-algebraic optimization*, Cambridge Texts in Applied Mathematics, Cambridge University Press, Cambridge, 2015. MR3469431

[61] Monique Laurent, *Sums of squares, moment matrices and optimization over polynomials*, Emerging applications of algebraic geometry, IMA Vol. Math. Appl., vol. 149, Springer, New York, 2009, pp. 157–270, DOI 10.1007/978-0-387-09686-5_7. MR2500468

[62] Benoît Legat, Raphaël M. Jungers, and Pablo A. Parrilo, *Generating unstable trajectories for switched systems via dual sum-of-squares techniques*, HSCC'16—Proceedings of the 19th International Conference on Hybrid Systems: Computation and Control, ACM, New York, 2016, pp. 51–60, DOI 10.1145/2883817.2883821. MR3578686

[63] Eunji Lim and Peter W. Glynn, *Consistency of multidimensional convex regression*, Oper. Res. **60** (2012), no. 1, 196–208, DOI 10.1287/opre.1110.1007. MR2911667

[64] Johan Löfberg, *Pre- and post-processing sum-of-squares programs in practice*, IEEE Trans. Automat. Control **54** (2009), no. 5, 1007–1011, DOI 10.1109/TAC.2009.2017144. MR2518113

[65] L. Lovász, *On the Shannon capacity of a graph*, IEEE Transactions on Information theory, **25**, no. 1, 1–7, 1979.

[66] A. M. Lyapunov, *The general problem of the stability of motion*, Taylor & Francis Group, London, 1992. Translated from Edouard Davaux's French translation (1907) of the 1892 Russian original and edited by A. T. Fuller; With an introduction and preface by Fuller, a biography of Lyapunov by V. I. Smirnov, and a bibliography of Lyapunov's works compiled by J. F. Barrett; Lyapunov centenary issue; Reprint of Internat. J. Control **55** (1992), no. 3 [MR1154209 (93e:01035)]; With a foreword by Ian Stewart. MR1229075

[67] A. Magnani, S. Lall, and S. Boyd, *Tractable fitting with convex polynomials via sum-of-squares*, Proceedings of the 44th IEEE Conference on Decision and Control, 2005, pages 1672–1677. IEEE.

[68] A. Majumdar, A. A. Ahmadi, and R. Tedrake, *Control design along trajectories with sums of squares programming*, Proceedings of the IEEE International Conference on Robotics and Automation, 2013.

[69] Rahul Mazumder, Arkopal Choudhury, Garud Iyengar, and Bodhisattva Sen, *A computational framework for multivariate convex regression and its variants*, J. Amer. Statist. Assoc. **114** (2019), no. 525, 318–331, DOI 10.1080/01621459.2017.1407771. MR3941257

[70] A. Megretski, *SPOT: systems polynomial optimization tools.* Available at `https://github.com/spot-toolbox/spotless`, 2013.

[71] Katta G. Murty and Santosh N. Kabadi, *Some NP-complete problems in quadratic and nonlinear programming*, Math. Programming **39** (1987), no. 2, 117–129, DOI 10.1007/BF02592948. MR916001

[72] Brendan O'Donoghue, Eric Chu, Neal Parikh, and Stephen Boyd, *Conic optimization via operator splitting and homogeneous self-dual embedding*, J. Optim. Theory Appl. **169** (2016), no. 3, 1042–1068, DOI 10.1007/s10957-016-0892-3. MR3501397

[73] Martin J. Osborne and Ariel Rubinstein, *A course in game theory*, MIT Press, Cambridge, MA, 1994. MR1301776

[74] A. Papachristodoulou and S. Prajna, *On the construction of Lyapunov functions using the sum of squares decomposition*, IEEE Conference on Decision and Control, 2002.

[75] A. Papachristodoulou and S. Prajna, *A tutorial on sum of squares techniques for systems analysis*, American Control Conference, 2005. Proceedings of the 2005, 2005, pages 2686–2700. IEEE.

[76] P. A. Parrilo, *Structured semidefinite programs and semialgebraic geometry methods in robustness and optimization*, PhD thesis, California Institute of Technology, May 2000.

[77] P. A. Parrilo, *Polynomial games and sum of squares optimization*, Proceedings of the 45th IEEE Conference on Decision and Control, 2006.

[78] Pablo A. Parrilo and Ali Jadbabaie, *Approximation of the joint spectral radius using sum of squares*, Linear Algebra Appl. **428** (2008), no. 10, 2385–2402, DOI 10.1016/j.laa.2007.12.027. MR2408034

[79] Pablo A. Parrilo and Ronen Peretz, *An inequality for circle packings proved by semidefinite programming*, Discrete Comput. Geom. **31** (2004), no. 3, 357–367, DOI 10.1007/s00454-003-2880-2. MR2036943

[80] Pablo A. Parrilo and Bernd Sturmfels, *Minimizing polynomial functions*, Algorithmic and quantitative real algebraic geometry (Piscataway, NJ, 2001), DIMACS Ser. Discrete Math. Theoret. Comput. Sci., vol. 60, Amer. Math. Soc., Providence, RI, 2003, pp. 83–99. MR1995016

[81] G. Pólya. *Über positive Darstellung von Polynomen*, Vierteljschr. Naturforsch. Ges. Zürich, **73**, 1928, 141–145.

[82] S. Prajna and A. Jadbabaie, *Safety verification of hybrid systems using barrier certificates*, Hybrid Systems: Computation and Control, 2004, Springer, pages 477–492.

[83] S. Prajna, A. Papachristodoulou, and P. A. Parrilo. *SOSTOOLS: Sum of squares optimization toolbox for MATLAB*, 2002-05. Available from `http://www.cds.caltech.edu/sostools` and `http://www.mit.edu/~parrilo/sostools`.

[84] F. Ramsey and D. Schafer, *Sleuth2: Data Sets from Ramsey and Schafer's "Statistical Sleuth (2nd Ed)"*, 2016. R package version 2.0-4.

[85] Annie Raymond, Mohit Singh, and Rekha R. Thomas, *Symmetry in Turán sums of squares polynomials from flag algebras*, Algebr. Comb. **1** (2018), no. 2, 249–274, DOI 10.5802/alco. MR3856524

[86] T. Roh, B. Dumitrescu, and L. Vandenberghe, *Multidimensional FIR filter design via trigonometric sum-of-squares optimization*, IEEE Journal of Selected Topics in Signal Processing, **1**, no. 4, 2007, 641–650.

[87] M. Roozbehani, E. Feron, and A. Megrestki, *Modeling, optimization and computation for software verification*, International Workshop on Hybrid Systems: Computation and Control, Springer, 2005, pages 606–622.

[88] Gian-Carlo Rota and Gilbert Strang, *A note on the joint spectral radius*, Nederl. Akad. Wetensch. Proc. Ser. A 63 = Indag. Math. **22** (1960), 379–381. MR0147922

[89] C. W. Scherer and C. W. J. Hol, *Matrix sum-of-squares relaxations for robust semi-definite programs*, Math. Program. **107** (2006), no. 1-2, Ser. B, 189–211, DOI 10.1007/s10107-005-0684-2. MR2218126

[90] Emilio Seijo and Bodhisattva Sen, *Nonparametric least squares estimation of a multivariate convex regression function*, Ann. Statist. **39** (2011), no. 3, 1633–1657, DOI 10.1214/10-AOS852. MR2850215

[91] P. Shah and P. A. Parrilo, *Polynomial stochastic games via sum of squares optimization*, Conference on Dynamics and Control, 2007.

[92] James E. Smith, *Generalized Chebychev inequalities: theory and applications in decision analysis*, Oper. Res. **43** (1995), no. 5, 807–825, DOI 10.1287/opre.43.5.807. MR1361341

[93] Noah D. Stein, Pablo A. Parrilo, and Asuman Ozdaglar, *Correlated equilibria in continuous games: characterization and computation*, Games Econom. Behav. **71** (2011), no. 2, 436–455, DOI 10.1016/j.geb.2010.04.004. MR2814881

[94] J. Sturm, *SeDuMi version 1.05*, Oct. 2001. Latest version available at `http://sedumi.ie.lehigh.edu/`.

[95] K. C. Toh, R. H. Tütüncü, and M. J. Todd, *SDPT3 - a MATLAB software package for semidefinite-quadratic-linear programming.* Available from `http://www.math.cmu.edu/~reha/sdpt3.html`.

[96] Frank Vallentin, *Symmetry in semidefinite programs*, Linear Algebra Appl. **430** (2009), no. 1, 360–369, DOI 10.1016/j.laa.2008.07.025. MR2460523

[97] Hayato Waki, Sunyoung Kim, Masakazu Kojima, and Masakazu Muramatsu, *Sums of squares and semidefinite program relaxations for polynomial optimization problems with structured sparsity*, SIAM J. Optim. **17** (2006), no. 1, 218–242, DOI 10.1137/050623802. MR2219151

[98] Tillmann Weisser, Jean B. Lasserre, and Kim-Chuan Toh, *Sparse-BSOS: a bounded degree SOS hierarchy for large scale polynomial optimization with sparsity*, Math. Program. Comput. **10** (2018), no. 1, 1–32, DOI 10.1007/s12532-017-0121-6. MR3773087

[99] W. Whitt, *On approximations for queues. I. Extremal distributions*, AT&T Bell Labs. Tech. J. **63** (1984), no. 1, 115–138. MR731797

[100] Makoto Yamashita, Katsuki Fujisawa, and Masakazu Kojima, *Implementation and evaluation of SDPA 6.0 (semidefinite programming algorithm 6.0)*, Optim. Methods Softw. **18** (2003), no. 4, 491–505, DOI 10.1080/1055678031000118482. The Second Japanese-Sino Optimization Meeting, Part II (Kyoto, 2002). MR2019042

[101] Liuqin Yang, Defeng Sun, and Kim-Chuan Toh, SDPNAL+*: a majorized semismooth Newton-CG augmented Lagrangian method for semidefinite programming with nonnegative constraints*, Math. Program. Comput. **7** (2015), no. 3, 331–366, DOI 10.1007/s12532-015-0082-6. MR3384939

[102] Yang Zheng, Giovanni Fantuzzi, Antonis Papachristodoulou, Paul Goulart, and Andrew Wynn, *Chordal decomposition in operator-splitting methods for sparse semidefinite programs*, Math. Program. **180** (2020), no. 1-2, Ser. A, 489–532, DOI 10.1007/s10107-019-01366-3. MR4062844

Department of Decision Sciences, INSEAD, Fontainebleau, 77300, France
Email address: georgina.hall@insead.edu

Index

Selected Published Titles in This Series

77 **Pablo A. Parrilo and Rekha R. Thomas, Editors,** Sum of Squares: Theory and Applications, 2020

75 **Michael Damron, Firas Rassoul-Agha, and Timo Seppäläinen, Editors,** Random Growth Models, 2018

74 **Jan Bouwe van den Berg and Jean-Philippe Lessard, Editors,** Rigorous Numerics in Dynamics, 2018

73 **Kasso A. Okoudjou, Editor,** Finite Frame Theory, 2016

72 **Van H. Vu, Editor,** Modern Aspects of Random Matrix Theory, 2014

71 **Samson Abramsky and Michael Mislove, Editors,** Mathematical Foundations of Information Flow, 2012

70 **Afra Zomorodian, Editor,** Advances in Applied and Computational Topology, 2012

69 **Karl Sigmund, Editor,** Evolutionary Game Dynamics, 2011

68 **Samuel J. Lomonaco, Jr., Editor,** Quantum Information Science and Its Contributions to Mathematics, 2010

67 **Eitan Tadmor, Jian-Guo Liu, and Athanasios E. Tzavaras, Editors,** Hyperbolic Problems: Theory, Numerics and Applications, 2009

66 **Dorothy Buck and Erica Flapan, Editors,** Applications of Knot Theory, 2009

65 **L. L. Bonilla, A. Carpio, J. M. Vega, and S. Venakides, Editors,** Recent Advances in Nonlinear Partial Differential Equations and Applications, 2007

64 **Reinhard C. Laubenbacher, Editor,** Modeling and Simulation of Biological Networks, 2007

63 **Gestur Ólafsson and Eric Todd Quinto, Editors,** The Radon Transform, Inverse Problems, and Tomography, 2006

62 **Paul Garrett and Daniel Lieman, Editors,** Public-Key Cryptography, 2005

61 **Serkan Hoşten, Jon Lee, and Rekha R. Thomas, Editors,** Trends in Optimization, 2004

60 **Susan G. Williams, Editor,** Symbolic Dynamics and its Applications, 2004

59 **James Sneyd, Editor,** An Introduction to Mathematical Modeling in Physiology, Cell Biology, and Immunology, 2002

58 **Samuel J. Lomonaco, Jr., Editor,** Quantum Computation, 2002

57 **David C. Heath and Glen Swindle, Editors,** Introduction to Mathematical Finance, 1999

56 **Jane Cronin and Robert E. O'Malley, Jr., Editors,** Analyzing Multiscale Phenomena Using Singular Perturbation Methods, 1999

55 **Frederick Hoffman, Editor,** Mathematical Aspects of Artificial Intelligence, 1998

54 **Renato Spigler and Stephanos Venakides, Editors,** Recent Advances in Partial Differential Equations, Venice 1996, 1998

53 **David A. Cox and Bernd Sturmfels, Editors,** Applications of Computational Algebraic Geometry, 1998

52 **V. Mandrekar and P. R.Masani, Editors,** Proceedings of the Norbert Wiener Centenary Congress, 1994, 1997

51 **Louis H. Kauffman, Editor,** The Interface of Knots and Physics, 1996

50 **Robert Calderbank, Editor,** Different Aspects of Coding Theory, 1995

49 **Robert L. Devaney, Editor,** Complex Dynamical Systems: The Mathematics Behind the Mandelbrot and Julia Sets, 1994

48 **Walter Gautschi, Editor,** Mathematics of Computation 1943–1993: A Half-Century of Computational Mathematics, 1995

For a complete list of titles in this series, visit the
AMS Bookstore at **www.ams.org/bookstore/psapmseries/**.